Galactic and Intergalactic Magnetic Fields

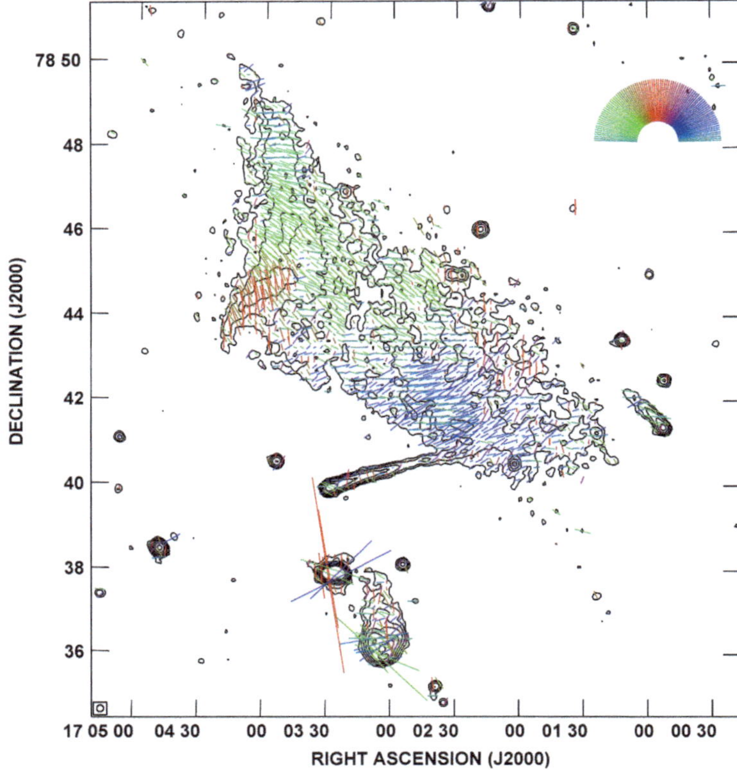

Radio synchrotron emission from the galaxy cluster A 2256 at $\lambda = 13$ cm, observed with the WSRT (M. Trasatti, Ph.D. thesis, Univ. Bonn, 2014). Contour lines depict the total intensity, the coloured bars show the orientation of the magnetic field, with their length proportional to the degree of linear polarisation.

More information about this series at
http://www.springer.com/series/4097

Ulrich Klein • Andrew Fletcher

Galactic and Intergalactic Magnetic Fields

 Springer

Ulrich Klein
Argelander-Institut für Astronomie
Universiät Bonn
Bonn
Germany

Andrew Fletcher
School of Mathematics and Statistics
Newcastle University
Newcastle upon Tyne
United Kingdom

ISBN 978-3-319-34671-7   ISBN 978-3-319-08942-3 (eBook)
DOI 10.1007/978-3-319-08942-3
Springer Cham Heidelberg New York Dordrecht London

Cover illustration: Radio synchrotron emission from the galaxy cluster A 2256 at $\lambda = 13\,\mathrm{cm}$, observed with the WSRT (M. Trasatti, Ph.D. thesis, Univ. Bonn, 2014). Contour lines depict the total intensity, the coloured bars show the orientation of the magnetic field, with their length proportional to the degree of linear polarisation.
Cover design: Jim Wilkie

Printed on acid-free paper

Springer is part of Springer Science+Business Media (www.springer.com)

Preface

This textbook arises from a lecture that has been created in spring 2010 for students of astronomy at Bonn University and was first held in the summer term of that year. The motivation for this course was threefold: first, magnetic fields in the interstellar and intergalactic medium had been a central research topic at Bonn (Max-Planck-Institut für Radioastronomie and Universität) for more than three decades, while the students involved in such projects had little or no education on this topic. Second, this topic was an obvious gap to be filled in an otherwise rather complete astrophysical curriculum. And last but not least, a so-called research unit entitled *Magnetisation of Interstellar and Intergalactic Media: The Prospects of Low-Frequency Radio Observations* was granted by the DFG (the German Research Foundation), for research projects of which interested students had to be recruited and educated.

The textbook presents a fairly thorough treatment of the main radiation process which the bulk of our knowledge about magnetic fields in the diffuse interstellar and intergalactic medium rests upon, namely synchrotron radiation. The spectrum of thermal free-free radiation is also derived, as in case of galaxies it contaminates the total radio emission that we utilise to explore the properties of magnetic fields. It then describes phenomena that have been observed with various tools at different wavelengths, with theoretical concepts addressed wherever necessary. It is meant to give an overview of our knowledge of various astrophysical phenomena related to cosmic magnetism, with scales ranging from molecular clouds in star-forming regions and supernova remnants in the Milky Way, to clusters of galaxies on the largest scales. The textbook is not dealing with magnetic fields in condensed matter (planets, stars, neutron stars). It shall be helpful for both Master and Ph.D. students who are interested in this subject and/or work in this field of research.

Finally, I would like to thank Prof. Dr. Richard Wielebinski: it was him who 'whetted my appetite' on this field of research back in the late 1970s.

Uli Klein

The details of dynamo theory are still being argued about, sometimes passionately! This is good for the field of research but can be hard on an interested newcomer: as well as the usual specialised terminology it quickly becomes difficult to avoid dealing with at least some of the mathematics, which can rapidly get complicated. The text about dynamos in this book is an attempt to provide an astrophysical and mathematical background that will allow the reader to understand the gist of current research and the debates which surround it. To that end, some of the ideas are presented in a qualitative way, rather than via a rigorous mathematical derivation. Hopefully a reader who feels short-changed by this approach will be able to satisfy their curiosity using the recommendations for further reading.

Some of the material is based on a course on instabilities that I have taught for several years in the School of Mathematics and Statistics at Newcastle University. This was originally devised by Prof. Carlo Barenghi and has benefited from the critical questions of several cohorts of third year mathematics students.

I am fortunate in having worked with Anvar Shukurov of Newcastle University and Rainer Beck and Elly Berkhuijsen of the Max-Planck-Institut für Radioastronomie in Bonn from the very beginning of my research career. Without this collaboration, and the friendly discussions with many others working in the field over many years, my enthusiasm would have waned a long time ago. Comments from Anvar Shukurov greatly improved the text on dynamos. Finally, Brigitta von Rekowski has my huge thanks for checking and correcting my mathematics and English!

Andrew Fletcher

It may be unusual that a textbook comes into existence over a socialising pint of beer. Actually the authors first met at Dublin in October 2010, during the 4th workshop of the LOFAR Key Science Project on Cosmic Magnetism. This may sound strange, but the first author had been working on a rather different field for the preceding two decades, and so he even missed to make the second author's acquaintance during his postdoc period at the MPIfR 'next door'. The two chaps quickly found out that they were giving lectures at Bonn and Newcastle, respectively, which are obviously complementary. After a few minutes, the irrevocable decision was made to put these lectures together for a textbook on a subject for which most information until now had to be searched in pertinent journals.

It is our hope that the interested reader is contented with the information conveyed by the book, and is inspired to read more using the references given in each chapter. The book shall mainly serve to gain knowledge about magnetic fields in diffuse astrophysical media and to understand the processes responsible for their creation and sustainment. Our concise text shall thus also raise or increase the curiosity for this subject.

Cheers!

<table>
<tr><td>Bonn, Germany</td><td align="right">Uli Klein</td></tr>
<tr><td>Newcastle upon Tyne, UK</td><td align="right">Andrew Fletcher</td></tr>
<tr><td>February 1, 2014</td><td></td></tr>
</table>

Contents

List of Figures

List of Tables

Chapter 1
Introduction

1.1 Magnetism

Historically, the phenomenon of magnetism was recognised in the form lodestones, which are natural magnetised pieces of iron ore. The term *magnet* in Greek meant "stone from Magnesia", a part of ancient Greece where lodestones were found. When lodestones were suspended such that they could turn they served as the first magnetic compasses. The earliest records of this phenomenon are from Greece, India, and China around 2500 years ago. Apart from iron, magnetism was exhibited in a natural state by cobalt and manganese, and by many of their compounds. The phenomenon was coined "magnetism" (derived from city of Magnesia in Asia Minor, where they say the phenomenon was first recognised).

The strength of magnetic fields is given in units of Gauss, and a frequently used unit is $1\,T = 10^4\,G$. In order to get a feeling for this magnetic force, a comparison with some weight may be helpful: holding a weight of 1 kg in the earth's gravitational field corresponds to force of 9.81 N. Keeping apart two bar magnets ($B = 50\,G$) that are separated by 0.4 cm requires this force.

Our earth can be considered as a huge magnet: a magnetised rod (i.e. a compass) suspended to rotate freely finally aligns with the earth's magnetic field, with the north pole of the compass pointing northwards, towards the geographic north pole, which is actually the magnetic south pole! The earth's magnetic field protects us against high-energy (charged) particles, the cosmic rays (CRs, see Chap. 4) by trapping them in the so-called van Allen radiation belts. This magnetic field has an overall dipolar morphology, with a field strength near to the earth's surface of $B_E \approx 0.3 \ldots 0.5\,G$. For comparison, a small iron magnet has 100 G, the magnet used in nuclear magnetic resonance imaging (NMRI) has between 5000 and 70000 G. Neutron stars represent the most extreme environment in this respect: they have $\leq 10^{15}\,G$ on their surface.

The magnetic field of the sun is of cardinal significance for many processes on its surface. For instance by heating the corona to its temperature of $\sim 10^6\,K$ via

© Springer International Publishing Switzerland 2015

U. Klein, A. Fletcher, *Galactic and Intergalactic Magnetic Fields*,

Springer Praxis Books, DOI 10.1007/978-3-319-08942-3_1

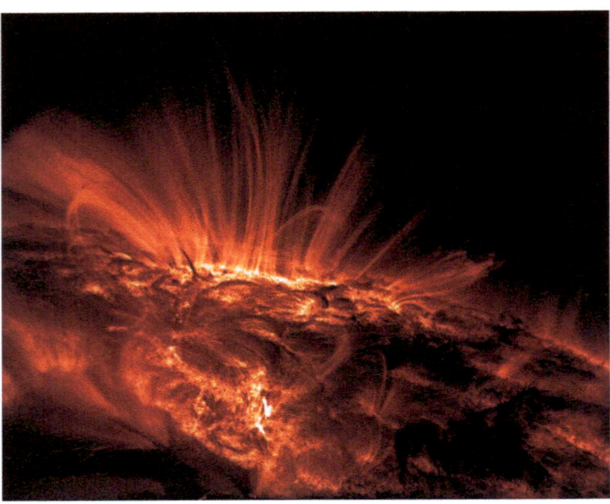

Fig. 1.1 Magnetic loops on the sun (Credits: Trace Project, NASA)

magnetic reconnection continuously accelerates the charged particles. The sunspots mark locations of an enhanced (0.1 T) magnetic field with specific polarity (N or S), which means that sunspots are bound to occur in pairs, connected by magnetic loops (Fig. 1.1). The polarity of the preceding and following bipolar spots (in the sense of the sun's rotation) is always the same in one (northern) solar hemisphere, and is opposite in the other (southern). This polarity is reversed every 22 years (solar cycle). We shall briefly come back to the solar magnetic field when discussing cosmic dynamos.

As the title of the book suggests, we shall not deal here with magnetic fields inside planets, stars or other more condensed celestial bodies, but rather with the magnetic fields between them, i.e. those pervading interstellar and intergalactic space. The reason for this is twofold; it is partly technical, as the techniques to detect or measure magnetic fields inside condensed celestial bodies are mostly different from those 'probing' magnetic fields in diffuse matter. This has also to do with the magnetic-field strengths, which in general reach but a few μG in diffuse cosmic media, while they are (many) orders of magnitude higher in planets, stars, or stellar relics. It turns out that the bulk of our knowledge about large-scale magnetic fields in the universe rests upon the observed properties of synchrotron radiation, mostly observed in the radio regime. This also justifies that we spend two chapters on the radiation mechanism and its diagnostics (not neglecting though other methods).

Magnetic fields exert a force on the cosmic plasma and, depending on the size scale, have a more or less strong influence on its motion. As we shall see, they can not compete with the overall gravitational forces at work in galaxies, let alone on even larger scales, but they have a strong influence on the gas motions on small scales. As an example, their role in controlling the collapse of molecular clouds preceding star formation has to be considered. For sure, their interaction with the relativistic gas component is anything but negligible. They control the propagation of cosmic rays impinging onto the earth, their streaming through the solar system, in the Milky Way as well as in external galaxies and in intergalactic space. Hence, understanding the evolution of the interstellar and intergalactic medium can not solely rest upon sheer hydrodynamics, but has to involve the phenomenon of magnetism! This, among other aspects, is the main message of this text book to students of astrophysics.

Throughout this book, we use cgs units (Gaussian system), which is quite common in astrophysics. In the context of electro-magnetism, this bears the advantage that we do not have to deal with the constants ϵ_0 and μ_0, so that e.g. the energy density of magnetic fields is simply $B^2/8\pi$ (instead of $B^2/2\mu_0$ in SI units; see also Sect. 6.3.1).

1.2   Magnetic Force on a Moving Charge

A charged particle (e.g an electron) moving in a magnetic field experiences a centripetal force that accelerates it perpendicular to its direction of motion (Fig. 1.2)

$$\vec{F} = \frac{e}{c}\,\vec{v} \times \vec{B}, \tag{1.1}$$

where $\vec{B}$ is the magnetic field strength, e the unit charge, c the speed of light, and v the velocity of the particle. This so-called Lorentz force forms a thread through this book, since it implies that as soon as we deal with charged particles, it is this force that controls their motion and thereby leads to a tight coupling between charged particles and magnetic fields. Most of the gas in the universe is ionised, and hence is a plasma. In galaxies, the UV and X-ray photons as well as the CRs keep the ionisation at some level, while in clusters of galaxies the gas is nearly completely ionised, owing to the energy input by large-scale shock waves. Hence, the interstellar and intergalactic matter is always ionised to some extent so that the gas and the magnetic fields are always strongly coupled and can therefore never be considered separately. This is a major challenge for any theoretical modelling of processes connected with gas flows.

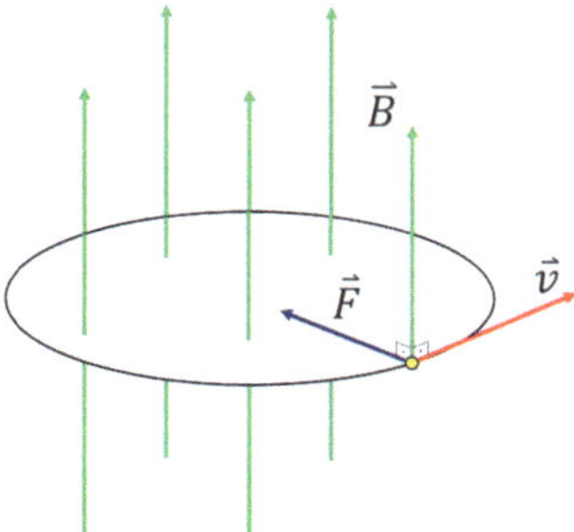

Fig. 1.2 Lorentz force

Equating (1.1) to the centrifugal force, one can calculate the so-called radius of gyration r_L, also called the Larmor radius. If $\vec{v} \perp \vec{B}$, we have

$$\frac{m\,v^2}{r_L} = e\,v\,\frac{B}{c} \tag{1.2}$$

and hence

$$r_L = \frac{m\,v\,c}{e\,B}. \tag{1.3}$$

With

$$v = \omega\,r_L,$$

the resulting gyration or Larmor frequency is

$$\omega = \frac{e}{m\,c}\,B, \tag{1.4}$$

1.3   Magnetisation of Matter

Atoms and molecules may or may not possess a net magnetic dipole moment, depending on their symmetry and on the relative orientation of their electronic orbits. Matter in bulk, with the exception of ferromagnetic substances, does not exhibit any net magnetic moment, owing to the random orientation of the atoms or molecules. However, when applying an external magnetic field, the distortion of the electronic motion results in a net magnetisation. Depending on the composition of the substance, we distinguish the following kinds of magnetism occurring in solid-state material (Fig. 1.3):

- Diamagnetism: This can be considered as a general property of any substance! When applying an external magnetic field to matter, this produces a magnetic dipole moment opposite to the one applied,[1] which is hence repulsive. This magnetisation, called "diamagnetism" is, however, masked by paramagnetism in most substances.
- Paramagnetism: Substances with unpaired electrons possess a permanent magnetic dipole moment, associated with the angular momentum of the unpaired electrons. Applying an external magnetic field tends to align all magnetic moments *along* the external field, which results in a magnetisation, called "paramagnetism". So here, the magnetisation is in the same direction as the

---

[1] In a strong laboratory $\vec{B}$-field (≤ 15 T), strongly diamagnetic substances can levitate!

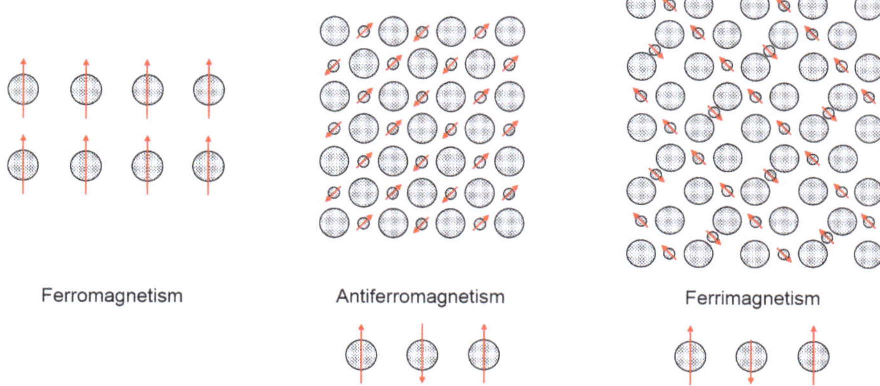

Fig. 1.3 Different kinds of magnetism in solid bodies

external magnetic field. It is much stronger than diamagnetism and is attractive rather than repelling.

- Ferromagnetism: This form of magnetism implies permanent magnetisation, due to the tendency of magnetic moments to align under their mutual magnets. It results in so-called "Weiss domains", their sizes being of order $10\,\mu$m...0.2 mm. This is the characteristic property of iron (Fe).
- Antiferromagnetism: In some substances, the electronic spins are antiparallel, which results in a vanishing magnetisation. Such substances are, for instance, MnO, FeO, CoC, and NiO.
- Ferrimagnetism: This kind of magnetism is similar to that of antiferromagnetism, but the atomic or ionic magnetic moments in one direction are different from those in the other. These substances are called "ferrites" and are compounds of the form $MOFe_2O_3$, where M stands for Mn, Co, NI, CU, Mg, Zn, Cd etc. If M=Fe, then the compound Fe_3O_4, or magnetite, results.

Concluding, we find that magnetism in solid-state matter is due to atomic structure and electronic orbits. In astrophysical plasmas, magnetism is due to large-scale electric currents, producing in particular the dynamo action (see Chap. 6). It is at work in the earth, in the sun, in planets, in the interstellar medium (ISM), and even in the intergalactic medium via hydrodynamic flows.

In every day's life, ferrites have a wide range of applications, as they have little or no electric conductivity:

- Iron cores of transformers (low eddy-current losses)
- High-frequency electronic components (e.g. directional couplers)
- Noise suppression chokes ("ferrite head" in computer cables)
- Erase heads in tape recorders, computer disks
- Coating of magnetic tapes
- Loudspeaker magnets

1.4 Some History in Astrophysics

Before diving into the subject of cosmic magnetism, it may be worth to briefly review the history of cosmic magnetism by summarizing the landmarks of the notion that magnetic fields are ubiquitously present in the universe.[2] Prior to World-War II it was noted (e.g. Alfvén, 1937) that magnetic fields in sun and stars or in interstellar space could have a strength and scale sufficient to accelerate charged particles up to energies as observed in cosmic-ray experiments, revealing protons, electrons, and α particles.

1946   The first discrete radio source Cyg A was discovered by Hey et al. (1946). They noticed a fluctuation of the signal, which turned out to be low-frequency scintillation. Cyg A is the proto-typical radio galaxy emitting intense synchrotron radiation (see Sect. 7.4).

1948   Bolton and Stanley (1948), using their "sea interferometer", determined the angular size of Cyg A to be $8'$. Smith measured its precise position in 1951, facilitating an optical identification with the Palomar 200 in. telescope by Baade. The redshift implied a distance of 230 Mpc, resulting in a stunning radio luminosity of $L_{radio} \approx 10^{45}$ erg s^{-1}! Later on, interferometric observations revealed a central source, two jets and lobes. Such objects with $L_{radio} \approx (10^3 \dots 10^4) \times L_{MW}$ were termed "radio galaxies".

1949   The detection of the Crab Nebula in the radio regime was reported by Bolton and Stanley (1949). Called Tau A, it was identified with the supernova explosion reported by Chinese astronomers in 1054. Such supernova remnants also produce relativistic particles emitting synchrotron radiation (Sect. 4.3).

1949   Alfvén et al. (1949), Richtmyer and Teller proposed that cosmic rays have a solar origin and are confined to solar surroundings by an interplanetary magnetic field. With the Larmor radius r_L given by Eq. (1.3), the particles with an inferred energy $E_{max} \approx 10^{14}$ eV must have Larmor radii of up to $r_L = 3 \cdot 10^{17}$ cm for a magnetic field strength $B \approx 10^{-6}$ G, inferred from other evidence. For particles with $E \approx 10^{20}$ eV, such a µG field would keep them within the Galactic disk, because the Larmor radius has a size scale comparable to that of the thickness of the Galactic disk. To date, we know that the highest-energy particles ($E \geq 10^{21}$ eV) must come from outside the Milky Way, simply because there aren't any conceivable acceleration mechanisms within the Galaxy capable of producing such energies.

1949   Fermi (1949) proposed that CRs fill the whole Galaxy, so they would not be confined to the solar system. He suspected that a magnetic field exists on still larger scales. He was aware of the high conductivity of the ISM and of its moving inhomogeneities that could serve as efficient accelerators (Sect. 4.4).

---

[2]The very first proof of the existence of magnetic fields outside the earth was made by G.E. Hale, who in 1908 discovered magnetic fields in the sunspots via the Zeeman effect.

1949   Hiltner (1949) delivered impressive evidence for the existence of interstellar magnetic fields, when he discovered the polarisation of starlight. This linear polarisation is explained in terms of the so-called Davis-Greenstein mechanism (Sect. 3.1).

1950   The galactic nonthermal radio emission was proposed to be "magneto-bremsstrahlung" (synchrotron radiation, see Sect. 2.3) by Alfvén and Herlofson (1950), by Kiepenheuer (1950), and later on also by Ginzburg (1953) and Shklovsky (1952). The discovery of linearly polarised radio emission in the Milky Way strongly supported this view (Westerhout et al., 1962; Wielebinski et al. 1962). All of this evidence suggested the presence of a ubiquitous magnetic field in the Galaxy.

1956   Imprints of magnetic fields were noticed in the morphologies of nebulae in the Milky Way, which exhibit filamentary structures. Shajn (1956) suggested that the elongation of diffuse nebulae (along galactic plane) is connected with a large-scale interstellar magnetic field.

1963   Schmidt (1963) identified optically the radio source $3C\,273$[3] with a star-like object at $z = 0.158$, $D_L = 750\,\mathrm{Mpc}$ (ΛCDM). The name "QSO" (quasi-stellar object, also: "quasar") was coined. Later on, a whole "zoo" of active galactic nuclei (AGN) was revealed (radio galaxies, Seyfert galaxies, QSO, BL Lac, . . . ; see Sect. 7.1).

1964   The first measurement of Faraday rotation was reported by Morris and Berge (1964) and (later on) by Berge and Seielstad (1967) against extragalactic polarised radio sources. These measurements were followed by measurements of pulsars (Manchester 1972, 1974). A magnetic-field strength of $B_{ISM} \approx 2 \ldots 3\,\mu\mathrm{G}$ was inferred for the *ordered field component* (there is also a random one that does not contribute to RM, see Sect. 3.3).

1968   The first detection of the Zeeman splitting in clouds of neutral hydrogen (HI) was made by Verschuur (1968). This discovery was of cardinal importance to directly prove the existence of magnetic fields in the ISM. The problem is, however, that the splitting of the lines in frequency is minute (see Sect. 3.4). Hence this discovery came late.

1971   The first models of a galactic dynamo were worked out by Parker (1971a,b), and by Vainshtein and Ruzmaikin (1971). The concept of a hydromagnetic dynamo to explain the origin and maintenance of magnetic fields of the sun and earth actually goes back to Larmor.

1978   This year marks the first detection of polarised radio emission from an external spiral galaxy, viz. M 31, by Beck et al. (1978). These activities have developed to a major research field (not only) in Bonn.

---

[3]The 3C catalogue, meaning the *Third Cambridge Catalogue of Radio Sources* was published in 1959.

1.5 Significance of Interstellar and Intergalactical Magnetic Fields

Magnetic fields may have a strong influence on the various astrophysical processes going on in the interstellar and intergalactic medium. Its relative importance is essentially determined by the energy density

$$u_{mag} = \frac{B^2}{8\,\pi}.$$ (1.5)

Below, we summarise the most important aspects in which magnetic fields may be significant and may even play an active role. For instance, Eq. (1.5) tells us that magnetic fields cannot have any influence on the dynamics of galaxies, while on small scales (cloud collapse and subsequent star formation) this is rather different.

- Coming back to the very first inferences from the notion of the presence of magnetic fields in the ISM, it is obvious that our understanding of the strength and morphology of magnetic fields is of paramount importance for our understanding of CR transport. In particular, we need to know whether they arrive isotropically as suggested by observations of CRs at energies up to $\approx 10^{14}$ eV. What about ultra-high energy CRs (UHECR)? Is there any preferred direction of their arrival, possibly connected with any nearby AGN?
- Some more general and fundamental questions to be addressed are: what is the origin of magnetic fields in the ISM and intra-cluster or intergalactic medium (ICM/IGM)? Were they initially injected by AGN or by starburst (dwarf?) galaxies, together with the relativistic particles? What kind of mechanisms maintain and amplifie magnetic fields?
- One of the most important issues that will be a thread through a large part of this book is the role of magnetic fields in galactic evolution. Surely, magnetic fields take an active role in star formation, as they serve to transport angular momentum outwards during the cloud collapse; otherwise, star formation would be strongly inhibited. When stars cease their lives, they exert a strong feed-back on the ISM via galactic outflows or winds. As we meanwhile know, magnetic fields are being dragged along with such outflows.
- Some more specific questions related to large-scale phenomena in clusters of galaxies that are hitherto not fully understood: what is the nature of radio relics (cosmological shocks) in galaxy clusters? What is the nature of radio haloes in galaxy clusters? Are they produced by secondary electrons, or rather by shocks produced in galactic wakes? Are intergalactic filaments magnetised?
- Perhaps the most fundamental question pertains to the origin of magnetic fields: have there been primordial seed fields? Did they come into existence soon after the Big Bang, before the recombination era, or in the course of cosmological structure formation?

The hope is that this textbook conveys basic knowledge about this field of research to the students, and raise their interest in this. As far as contemporary knowledge allows, it will try to answer some of the questions addressed above. We presume that the reader is familiar with general astrophysics, i.e. the properties of the interstellar medium (ISM), the structure and main constituents of galaxies of various types, the properties of the intergalactic medium, and the large-scale structure of the universe and its evolution. These aspects become important as we explore the properties of magnetic fields from smaller scales within the Milky Way to large ones beyond the size scales of galaxies. An important point is that essentially the whole universe is a plasma. Stars are a perfect plasma, and so is the hot phase of the ISM, and even though the bulk of the ISM is cold and neutral, also these components are partially ionised and thus tightly coupled to the interstellar magnetic field. The hot gas in galaxy clusters which accounts for some 15 % of their total mass is again a perfect plasma. To conclude, if magnetic fields are a ubiquitous ingredient to cosmic structures, and there is good reason for this to hold true, magnetic fields can never be neglected, and their possible role in the evolution of these structures needs to be explored.

References

Alfvén, H.O.G., 1937, Nature **139**, 245

Alfvén, H.O.G., Richtmyer, Teller, E., 1949, Phys. Rev. **75**, 892

Alfvén, H.O.G., Herlofson, M., 1950, Phys. Rev. **78**, 616

Beck, R., Berkhuijsen, E.M., Wielebinski, R., 1978, Astron. Astroph. **68**, 27

Berge, G.L., Seielstad, G.A., 1967, Astroph. J. **148**, 367

Bolton, J.G., Stanley, G.J., 1948, Nature **161**, 312

Bolton, J.G., Stanley, G.J., 1949, Austr. J. Scient. Research. A. **2**, 139

Fermi, E., 1949, Phys. Rev. **75**, 1169

Ginzburg, V.L., 1953, Usp. Fiz. Nauk, **51**, 343. Fortschr. Phys., Berlin **1**, 659, 1954

Hey, J.S., Phillips, J.W., Parsons, S.J., 1946, Nature **157**, 296

Hiltner, W.A., 1949, Nature **163**, 283

Kiepenheuer, K.O., 1950, Phys. Rev. **79**, 738

Manchester, R.N., 1972, Astroph. J. **172**, 43

Manchester, R.N., 1974, Astroph. J. **188**, 63

Morris, D., Berge, G.L., 1964, Astroph. J. **139**, 1388

Parker, E.N., 1971a, Astroph. J. **163**, 279

Parker, E.N., 1971b, Astroph. J. **163**, 255

Schmidt, M., 1963, Nature **197**, 1040

Shklovsky, I.S., 1952, Astron. Zh. 29, 418

Shajn, G., 1956, Vist. Astron. **2**, 1066

Vainshtein, S.I., Ruzmaikin, A.A., 1971, Astron. Zh. **48**, 902

Verschuur, G.L., 1968, The Observatory, **88**, 115

Westerhout, G., Brouw, W.N., Muller, C.A., Tinbergen, J., 1962, Astron. J. 67, 590

Wielebinski, R., Shakeshaft, J.R., Pauliny-Toth, I.I.K., 1962, The Observatory, **82**, 158

Chapter 2
Continuum Radiation Processes

2.1 Radiation of an Accelerated Electron

After the very first inference that the nature of the Galactic low-frequency radiation must be synchrotron radiation by relativistic electrons moving in a magnetic field for which there is other independent evidence (Sect. 1.4), this radiation turns out to be the main diagnostic tool to trace magnetic fields in interstellar and intergalactic space, by virtue of its radiation spectrum and polarisation characteristics. It is therefore mandatory to work out these spectral and polarisation properties from first principles of electro-magnetic theory. The radiation is naturally measured at radio frequencies as shall be worked out below. However, this is not the only radiation process measured at centrimetric and decametric wavelengths, since the radiation of the ionised gas present in HII regions and in the diffuse ionised medium of the Milky Way and external galaxies contaminates the overall radio emission. Hence, we also need to understand the properties of this thermal emission, so we have to explore it here as well. We shall therefore in what follows work out the continuum radiation spectra of both, the thermal (free-free) and nonthermal (synchrotron) emission. The former is called 'thermal', as it results from an ensemble of particles with a Maxwellian energy distribution, while the energy distribution of the latter follows a power-law.

Of course, there is a multitude of textbooks in which one finds comprehensive treatments of continuum radiation processes, but an elaborate treatment in the context of this textbook appears quite natural, and the reader does not need to switch to another book in order to comprehend the essentials of these radiation mechanisms. As an example, one of the important properties of relativistic particles as opposed to thermal ones is that it takes orders of magnitude fewer of them to produce measurable synchrotron radiation, just because they are so energetic. We shall see that the relativistic component in a galaxy is totally insignificant by mass, yet it renders galaxies rather luminous in the radio domain.

© Springer International Publishing Switzerland 2015
U. Klein, A. Fletcher, *Galactic and Intergalactic Magnetic Fields*,
Springer Praxis Books, DOI 10.1007/978-3-319-08942-3_2

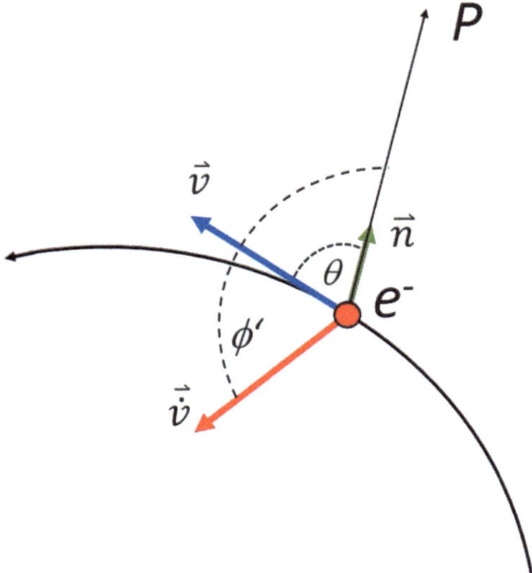

Fig. 2.1 Geometry for a moving charged particle as seen from some point P

Let us consider the electric field of an accelerated electron, with velocity $\vec{v}$ and acceleration $\dot{\vec{v}}$ as seen by an observer from some point $\mathcal{P}$ (Fig. 2.1). According to electro-magnetic theory[1] it is given by

$$\vec{E} = \frac{e}{c} \cdot \frac{\vec{n} \times [(\vec{n} - \vec{\beta}) \times \dot{\vec{\beta}}]}{R\,(1 - \cos\theta \cdot \beta)^3},\tag{2.1}$$

where $\vec{n}$ is the unit vector pointing from the particle towards the observer and

$$\vec{\beta} = \frac{\vec{v}}{c}, \ \dot{\vec{\beta}} = \frac{\dot{\vec{v}}}{c}.$$

The above expression contains the velocity $\vec{v}$ and the acceleration $\dot{\vec{v}}$. As we shall see, the radiated power of an accelerated charged particle is always proportional to $\dot{v}^2$, no matter whether it is thermal or relativistic. The flux of radiation is given by the Poynting vector

$$\vec{S} = \frac{c}{4\pi} \cdot \vec{E} \times \vec{B} = \frac{c}{4\pi} \cdot |\vec{E}^2| \cdot \vec{n}.\tag{2.2}$$

The power radiated into a unit solid angle per unit frequency and unit time is

---

[1]See, e.g. J.D. Jackson, *Classical Electrodynamics*, Chapt. 14.

$$\frac{dP(t)}{d\Omega} = |\vec{S}| \cdot (1 - \beta \cos\theta)\, R^2 = \frac{e^2}{4\,\pi\,c} \cdot \frac{|\vec{n} \times [(\vec{n} - \vec{\beta}) \times \dot{\vec{\beta}}]|^2}{(1 - \beta \cdot \cos\theta)^5}. \qquad (2.3)$$

This equation will be used for the case of thermal radiation, where $\beta \ll 1$, and non-thermal radiation, where $\beta \leq 1$ or $\gamma \gg 1$, by integrating the radiated power over the total $4\,\pi$ solid angle of the sphere. The angle $\theta = \angle(\vec{v}, \vec{n})$, hence

$$\cos\theta = \vec{n} \cdot \vec{\beta}.$$

R is the distance of point P between the electron and the observer. When measuring flux densities of radio sources, we can calculate their radio power or luminosity once we can determine this distance using standard astronomical techniques. Hence R is not relevant in the derivations that we shall work out below. It is just a matter of conversion from flux density to power or monochromatic luminosity, or from flux to total power or luminosity. So, converting for instance flux density S_ν to power P_ν then reads

$$P_\nu = 4\,\pi\,R^2\,S_\nu.$$

This, of course, presumes that the radio source emits isotropically, an assumption that is justified in most cases.

2.2 Free-Free Radiation

2.2.1 Physical Conditions in HII Regions

In an HII region, the hydrogen (and helium) atoms are predominantly ionised, with the free electrons moving on hyperbolic orbits (s.b.) about the protons (or ionised helium atoms). This is why the resulting radiation is termed 'free-free radiation' (in contrast to 'free-bound' or 'bound-free'). The electrons possess thermal velocities with a Maxwellian distribution, according to the temperature of the electron gas. In order to work out the radiated power of such a region as a function of frequency, we first need to compute the emitted power of a single accelerated electron.

This power is determined by the physical conditions of the plasma, viz. its density and temperature. Consider a plasma (HII region) with a temperature $T = 10^4$ K, which corresponds to a kinetic energy of the electrons of ≈ 1 eV. Let us assume that we are dealing mostly with protons and electrons (there will be some He ions, for which one can correct the results later on). The number densities, which are roughly identical for the electrons and protons, i.e. $n_e = n_p$ may range from 0.03 cm^{-3} in the general diffuse ISM, to 10^3 cm^{-3} in the centres of HII regions such as in the Orion Nebula (the radio source also called 'Ori A'). Due to their hyperbolic orbits about the protons the electrons permanently experience an acceleration, the main

Fig. 2.2 Trajectory of a
thermal electron in a plasma

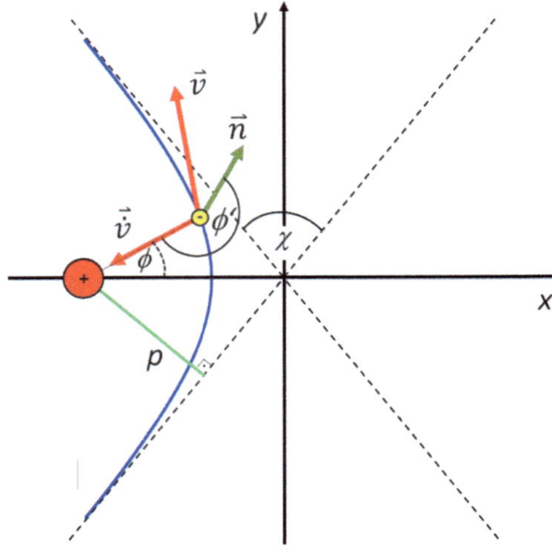

Fig. 2.3 Acceleration of an
electron

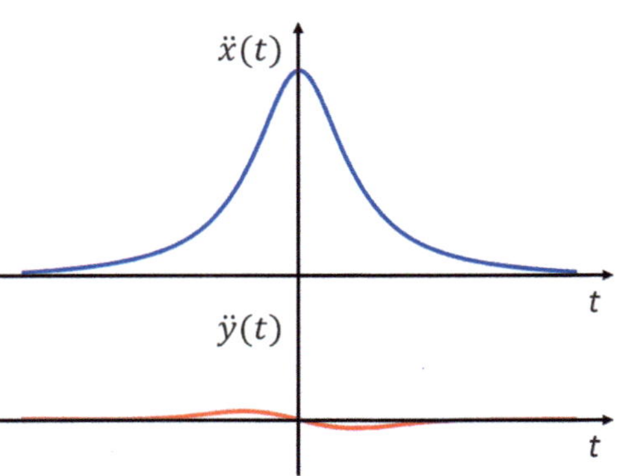

component of which is in the x-direction (Fig. 2.2). The motion is determined by
the Coulomb law (m_e: electron mass, Z: charge number, e: unit charge)

$$\dot{\vec{v}}(t) = -\frac{Z\,e^2}{m_e} \cdot \frac{\vec{r}}{r^3},$$

which has components (Fig. 2.3)

$$\ddot{x} = \frac{Z\,e^2}{m_e\,r^2} \cdot \cos\phi$$

and

$$\ddot{y} = \frac{Z\,e^2}{m_e\,r^2} \cdot \sin\phi.$$

2.2.2 *Radiation in the Non-relativistic Case*

We consider the non-relativistic case, for which $\beta \approx 0$, i.e. $(1 - \beta \cos \theta) = 1$. Hence, we have

$$\frac{dP}{d\Omega} = \frac{e^2}{4\pi c} \cdot \frac{|\vec{n} \times [(\vec{n} - \vec{\beta}) \times \dot{\vec{\beta}}]|^2}{(1 - \beta \cdot \cos \theta)^5} \tag{2.4}$$

$$= \frac{e^2}{4\pi c^3} \cdot \dot{v}^2(t) \, \sin^2 \phi', \tag{2.5}$$

and the integration over the solid angle yields the power radiated by a single electron as a function of time:

$$P(t) = \int_0^{2\pi} \int_0^{\pi} \frac{dP}{d\Omega} \cdot d\Omega = \frac{2\,e^2}{3\,c^3} \cdot \dot{v}^2(t) \tag{2.6}$$

since

$$\int_0^{2\pi} \int_0^{\pi} \sin^3 \phi' \, d\phi' \, d\theta = \frac{8\pi}{3}.$$

At this point, the various angles used above need to be explained once more: $\phi' = \angle(\vec{n}, \dot{\vec{\beta}})$, $\theta = \angle(\vec{n}, \vec{\beta})$, $\phi = \angle(\vec{x}, \dot{\vec{\beta}})$ (see Figs. 2.4 and 2.3).

We now know the radiated power as a function of time, $P(t)$, but we need to know the power as a function of frequency, $P(\nu)$, or $P(\omega)$, from which we can then derive the frequency spectrum of a single particle. The usual procedure to facilitate this conversion is to perform a Fourier analysis, or to calculate the Fourier transform

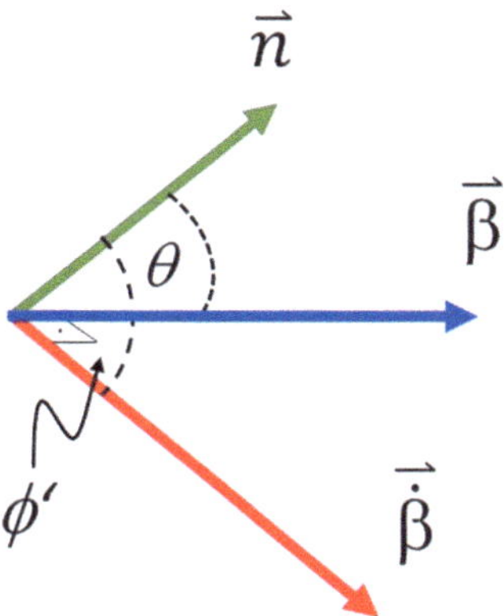

Fig. 2.4 Definition of angles

of the time-dependent quantity. We can thus express the time-dependent acceleration $\ddot{x}(t)$ in terms of its frequency-dependent Fourier transform

$$\ddot{x}(t) = \int_0^\infty C(\omega) \cdot \cos(\omega\, t) \cdot d\omega, \tag{2.7}$$

and vice versa

$$C(\omega) = \frac{1}{\pi} \int_{-\infty}^{+\infty} \ddot{x}(t) \cdot \cos \omega\, t \cdot dt. \tag{2.8}$$

The function $\ddot{x}(t)$ looks like a 'spike' as we have seen (Fig. 2.3). If it were a pulse (a δ-function), then it would have an infinitely wide frequency spectrum with constant power. Its finite width, however, gives rise to a finite frequency spectrum, dropping to zero at some critical frequency (Fig. 2.5). We can now evaluate $C(\omega)$ by performing the integration (2.8). We can utilise the modified Coulomb law to replace the integration over time t by an integration over the angle ϕ, which turns out to be trivial. We rewrite Coulomb's law, neglecting the y-term of the acceleration:

$$\ddot{x} = \frac{Z\, e^2}{m_e\, r^2} \cdot \cos \phi.$$

The particles move a distance $v \cdot dt$ during an infinitesimally small time step dt

$$v \cdot dt = d(p \cdot \tan \phi) = p\, \frac{d\phi}{\cos^2 \phi},$$

where we have introduced the impact parameter p. The above equation holds because p does not change significantly while $\ddot{x}$ is large, i.e. $dp = 0$. We can now

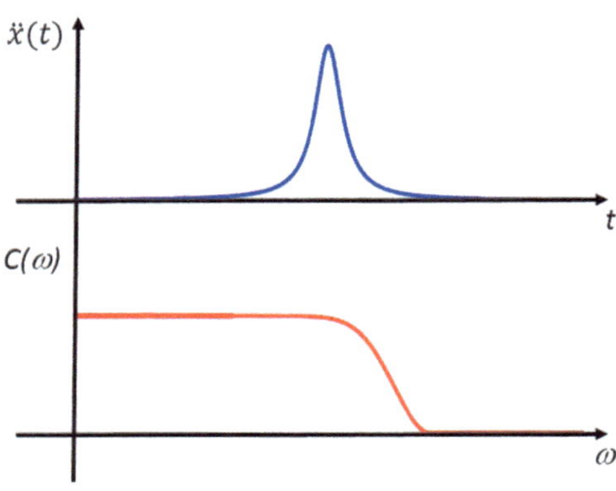

Fig. 2.5 Relation between $\ddot{x}(t)$ and $C(\omega)$

write the time step dt in terms of the angular increment $d\phi$,

$$dt = \frac{p}{v} \cdot \frac{d\phi}{cos^2\phi}.$$

The radiation will be significant only for the time interval

$$-\frac{p}{v} < t < \frac{p}{v},$$

for which $\omega t \ll 1$, hence $\cos \omega t \approx 1$. With $p = r \cdot \cos\phi$, and noting that the integration over ϕ can be reduced to the range between $-90°$ and $+90°$, we have

$$C(\omega) = \frac{1}{\pi} \int_{-\frac{\pi}{2}}^{+\frac{\pi}{2}} \frac{Z\,e^2 \cdot \cos^3\phi}{m_e\,p^2} \cdot \frac{p}{v} \cdot \frac{d\phi}{\cos^2\phi} \tag{2.9}$$

$$= \frac{Z\,e^2}{\pi\,m_e\,p\,v} \cdot \int_{-\frac{\pi}{2}}^{+\frac{\pi}{2}} \cos\phi\,d\phi \tag{2.10}$$

$$= \frac{2\,Z\,e^2}{\pi\,m_e\,p\,v}. \tag{2.11}$$

We have managed to invert the problem such that the function C does not depend on ω anymore. It has rather become a step function that approximates the real behaviour as sketched in Fig. 2.6. At this point, Parseval's theorem, which says that the integral over the radiated power must be the same in both domains (time and frequency) comes into play, as in the end we need to compute the radiated *power*! As $P(t) \sim \ddot{x}^2$ (Eq. 2.5),

$$\int_{-\infty}^{+\infty} \dot{v}^2(t)\,dt = \pi \int_{0}^{\infty} C^2(\omega)\,d\omega$$

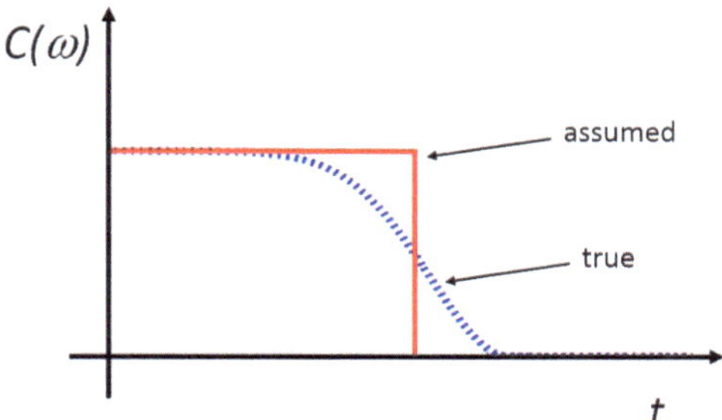

Fig. 2.6 Approximation of $C(\omega)$

$$\int\limits_{-\infty}^{+\infty} P(t)\, dt = \int\limits_{0}^{\infty} P(\omega)\, d\omega.$$

Using (2.5) this yields

$$P(\omega)\, d\omega = P(t)\, dt = \frac{2\, e^2}{3\, c^3} \cdot \pi\, C^2(\omega)\, d\omega$$

and, plugging in (2.11) from above, we obtain

$$P(\omega)\, d\omega = \frac{16\, e^6\, Z^2}{3\, m_e^2\, p^2\, v^2\, c^3}\, dv.$$

Recalling that $P(\omega) \approx 0$ for $v > v/p$ we have

$$P(v)\, dv = \frac{16\, e^6\, Z^2}{3\, m_e^2\, p^2\, v^2\, c^3}\, dv \qquad \text{for} \qquad v \le \frac{v}{p}$$

$$P(v)\, dv = 0 \qquad\qquad\qquad \text{for} \qquad v \ge \frac{v}{p}.$$

The latter of the above two equations tells us that only for infinitely small values of the impact parameter p would the frequency spectrum be constant out to infinity. We can make an estimate of the expected maximum frequency. Let us assume that we have volume densities $n_e = n_i = 10^3\,\mathrm{cm}^{-3}$ of the plasma, which corresponds to the conditions in the centre of an HII region. Let the electron temperature (definition s.b.) $T_e = 10^4\,\mathrm{K}$, which is typical. As we shall see below, this temperature corresponds to a mean velocity of the radiating electrons of $\bar{v} = 600\,\mathrm{km\,s}^{-1}$. The mean density implies a mean separation of the ions of $\bar{d} = 0.1\,\mathrm{cm}$, hence a mean impact parameter $\bar{p} \le 0.05\,\mathrm{cm}$, resulting in a frequency of $v \approx 1.2\,\mathrm{GHz}$. This is measured then in the radio regime!

2.2.3 Total Radiation

In order to calculate the power emitted by a certain ionised volume, we now consider the emission of an ensemble of thermal electrons. Let n_i be the number density of ions, n_e that of the electrons.

Let an ensemble of electrons be characterised by collision parameters between p and $p + dp$ (Fig. 2.7). Then the number of Coloumb collisions with this range of impact parameters $(p, p + dp)$ per second of time that an electron with velocity v will experience is equal to the number of ions within a cylindrical ring with length (per unit time) v and, respectively, inner and outer radii p and $p + dp$. This produces

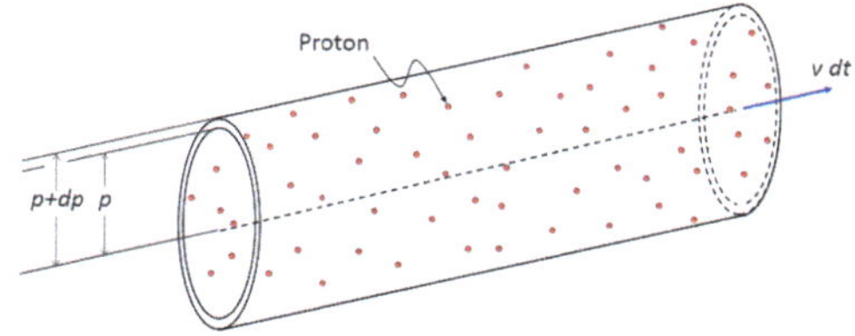

Fig. 2.7 Ensemble of protons, through which electrons with the impact parameter in the range (p, $p + dp$) are moving with speed v

$N_c = v \cdot 2\,\pi p\,dp \cdot n_i$ collisions each second. The number of collisions per cm^3 involving electrons with velocities between v and $v + dv$ then is

$$dN(v, p) = (v \cdot 2\pi p \cdot dp) \cdot n_i\, n_e \cdot f(v)\, dv \ \mathrm{cm}^{-3}\, \mathrm{s}^{-1}.$$

Here, $f(v)$ is the Maxwellian velocity distribution

$$f(v)\, dv = 4\pi \left(\frac{m_e}{2\,\pi\,k\,T_e} \right)^{\frac{3}{2}} e^{-\frac{m_e\, v^2}{2\,k\,T_e}}\, v^2\, dv, \tag{2.12}$$

where T_e is the so-called electron temperature. The total energy emitted per second, per Hz and per cm^3 is given by the emission coefficient ϵ_ν, integrated over $4\,\pi$ steradian:

$$
\begin{aligned}
4\,\pi\,\epsilon_\nu &= \int_{p_1}^{p_2} \int_0^\infty P(v, p)\, dN(v, p) \\[2ex]
&= \int_{p_1}^{p_2} \int_0^\infty \frac{16\, e^6 Z^2}{3\, c^3\, m_e^2\, p^2\, v^2} \cdot n_i\, n_e \cdot f(v) \cdot 2\,\pi\, p\, dp\, v\, dv \\[2ex]
&= \frac{32\,\pi\, e^6\, Z^2\, n_i\, n_e}{3\, c^3\, m_e^2} \cdot \int_0^\infty \frac{f(v)}{v}\, dv \cdot \int_{p_1}^{p_2} \frac{dp}{p}
\end{aligned}
$$

so that

$$\epsilon_\nu = \frac{8\, e^6\, Z^2 \cdot n_i\, n_e}{3\, c^3\, m_e^2} \cdot \int_0^\infty \frac{f(v)}{v}\, dv \cdot \int_{p_1}^{p_2} \frac{dp}{p}.$$

Note that the dimension of the emission coefficient is $[\epsilon_\nu] = \mathrm{erg\, s}^{-1}\, \mathrm{cm}^{-3}\, \mathrm{Hz}^{-1}\, \mathrm{sr}^{-1}$. The first integral is nothing but the mean reciprocal speed of the electrons, which upon plugging in the Maxwellian distribution (2.12) becomes

$$\int\limits_{0}^{\infty} \frac{f(v)}{v}\, dv = \left\langle \frac{1}{v} \right\rangle = \sqrt{\frac{2\, m_e}{\pi\, k\, T_e}},$$

where T_e is the electron temperature, with a typical value of $T_e = 10^4\,\mathrm{K}$. The second integral is called 'Gaunt factor' for the free-free radiation,

$$\int\limits_{p_1}^{p_2} \frac{dp}{p} = \ln \frac{p_2}{p_1} = g_{f\!f}. \tag{2.13}$$

This finally leads to an emission coefficient of the form

$$\epsilon_v = \frac{8\, e^6\, Z^2\, n_i\, n_e}{c^3\, m_e^{3/2}} \cdot \sqrt{\frac{2}{\pi\, k\, T_e}} \cdot g_{f\!f}.$$

We now need to discuss the most likely limits p_1 and p_2 for the impact parameter, which can be estimated as follows: the largest deflection that may occur, which implies the strongest acceleration and will thus produce the strongest radiation, is obviously connected with the smallest impact parameter. The smallest deflection that will still produce significant radiation corresponds to the largest relevant impact parameter.

Lower bound: The lower bound can be estimated in the following way. The electrons move on hyperbolic paths (Fig. 2.8), for which energy conservation holds, i.e.

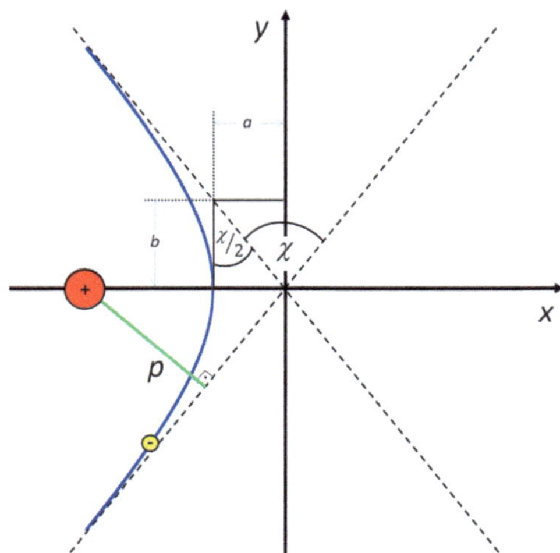

Fig. 2.8 Geometry of the hyperbolic orbit of the electron

$$E = E_{kin} + E_{pot} = \frac{1}{2} m_e v^2 - \frac{Z e^2}{r} = \frac{1}{2} m_e v_\infty^2 > 0$$
$$\overset{!}{=} \frac{Z e^2}{2 a}.$$

We can now simply set the minimum impact parameter p_1, by considering the relation between the deflection angle and the mass, charge, speed and impact parameter of the particle (see textbooks on classical mechanics):

$$\tan \frac{\chi}{2} = \frac{a}{b} = \frac{Z e^2}{b \, m_e \, v_\infty^2}.$$

The maximum deflection angle relevant here is $\chi_{max} = 90°$. Setting the corresponding minimum impact parameter equal to b, we obtain

$$\tan \frac{\chi_{max}}{2} = \tan \frac{\pi}{4} = \frac{Z e^2}{p_1 \, m_e \, v_\infty^2}.$$

From kinetic gas theory, we know that

$$\frac{1}{2} \cdot m_e \, v^2 = \frac{3}{2} \cdot k \, T_e$$

so that we can solve for p_1 to finally arrive at

$$p_1 = \frac{Z e^2}{3 \, k \, T_e}.$$

We could think of still smaller impact parameters, hence even lower bounds on p, such as the de-Broglie wavelength, but we would then leave the regime of classical mechanics and would have to apply relativistic corrections. For normal HII regions, this is not necessary, as their electron temperature is too low for this to be relevant. Such considerations would only have to be made in case of very hot, X-ray-emitting gas.

Upper bound: In order to calculate the upper bound p_2, we can argue that there must be a minimum deflection angle causing significant radiation (otherwise, there would not be sufficient acceleration to produce any radiation). Since this will correspond to the largest collision parameter, we can relate p_2 to the emitted frequency, corresponding to the inverse collision time:

$$2\pi \nu = \frac{1}{\tau_{coll}} = \frac{\langle v \rangle}{p_2}$$

so that

$$p_2 = \frac{\langle v \rangle}{2\pi v}$$

This time, we need to evaluate the mean speed of the electrons possessing a Maxwellian velocity distribution, which upon plugging in (2.12) becomes

$$\langle v \rangle = \int_0^\infty f(v)\, v\, dv = \sqrt{\frac{8\,k\,T_e}{\pi\,m_e}},$$

and hence

$$p_2 = \frac{1}{2\,\pi\,v} \cdot \sqrt{\frac{8\,k\,T_e}{\pi\,m_e}},$$

Thus, the Gaunt factor for free-free radiation becomes

$$g_{\mathit{ff}} = \ln \frac{3\,k\,T_e}{Z\,e^2} \cdot \frac{1}{2\,\pi\,v} \sqrt{\frac{8\,k\,T_e}{\pi\,m_e}}$$

$$= 12.5 + \ln\left[Z^{-1} \cdot \left(\frac{T_e}{10^4\,\mathrm{K}}\right)^{1.5} \cdot \left(\frac{v}{\mathrm{GHz}}\right)^{-1} \right] \approx 10\dots17,$$

which weakly depends on frequency and on the electron temperature. We now utilise Kirchhoff's law to obtain the absorption coefficient. This law relates the emission and absorption coefficients ϵ_v and $\varkappa_v$ via the so-called source function

$$S_v(T_e) = \frac{\epsilon_v}{\varkappa_v}.$$

In case of local thermo-dynamical equilibrium, the source function equals Planck's law,

$$B_v(T) = \frac{2\,h \cdot v^3}{c^2} \cdot \frac{1}{e^{\frac{h\,v}{k\,T_e}} - 1} \tag{2.14}$$

$$\approx \frac{2\,h\,v^2}{c^2} \cdot k\,T_e \quad \text{for} \quad \frac{h\,v}{k\,T_e} \ll 1. \tag{2.15}$$

In the radio domain, the Rayleigh-Jeans approximation (2.15) always holds, as the reader may easily verify. As an example of the situation discussed here, let us plug in an observing frequency $v = 700\,\mathrm{MHz}$ and an electron temperature $T_e = 10^4\,\mathrm{K}$, in which case we have $h\,v/k\,T_e \approx 3 \cdot 10^{-6}$. The absorption coefficient $\varkappa_v$, defined per unit path length, can now be computed from (2.14) and (2.15) to yield

$$\varkappa_\nu = \sqrt{\frac{2}{\pi}} \, \frac{4 \, e^6 \, Z^2 \, n_i \, n_e}{3 \, c \, (m_e \, k \, T_e)^{\frac{3}{2}} \, \nu^2} \cdot g_{ff}$$

$$= 8.77 \, 10^{-3} \left(\frac{n_i}{\mathrm{cm}^{-3}}\right) \left(\frac{n_e}{\mathrm{cm}^{-3}}\right) \left(\frac{\nu}{\mathrm{Hz}}\right)^{-2} \left(\frac{T_e}{\mathrm{K}}\right)^{-\frac{3}{2}} \mathrm{cm}^{-1}.$$

Next we compute the optical depth, which is a dimensionless quantity obtained by integrating the absorption coefficient over the total line-of-sight:

$$\boxed{\tau_\nu = \int_0^{s_0} \varkappa_\nu \, ds \qquad (2.16)}$$

so that

$$\tau_\nu = 8.77 \, 10^{-3} \left(\frac{\nu}{\mathrm{Hz}}\right)^{-2} \left(\frac{T_e}{\mathrm{K}}\right)^{-\frac{3}{2}} g_{ff} \cdot \int_0^{s_0} \left(\frac{n_i}{\mathrm{cm}^{-3}}\right) \left(\frac{n_e}{\mathrm{cm}^{-3}}\right) \left(\frac{ds}{\mathrm{pc}}\right).$$

In general the plasma is neutral, as there is zero net charge over the whole volume, i.e. we have $n_i = n_e$. At this point it is convenient to define a quantity called *emission measure*, usually designated as *EM*, in the following way

$$\boxed{EM = \int_0^{s_0} \left(\frac{n_e}{\mathrm{cm}^{-3}}\right)^2 \left(\frac{ds}{\mathrm{pc}}\right) \ \mathrm{pc} \ \mathrm{cm}^{-6}. \qquad (2.17)}$$

With this definition, which makes use of convenient units, we can now express the optical depth in the form

$$\tau_\nu = 3.01 \cdot 10^{-2} \cdot \left(\frac{EM}{10^6 \, \mathrm{pc} \, \mathrm{cm}^{-6}}\right) \cdot \left(\frac{\nu}{\mathrm{GHz}}\right)^{-2} \cdot \left(\frac{T_e}{10^4 \, \mathrm{K}}\right)^{-1.5} \cdot g_{ff},$$

again with convenient units. As an example, the Orion Nebula has an emission measure $EM \approx 10^6 \, \mathrm{pc} \, \mathrm{cm}^{-6}$ in its central region. The optical depth for free-free radiation becomes unity at a frequency of $\nu \approx 0.6 \, \mathrm{GHz}$ (with a Gaunt factor $g_{eff} = 12.3$ and an electron temperature $T_e = 10^4 \, \mathrm{K}$).

In order to account for the optical thickness, we need to employ a calculation of the radiative transfer. It allows us to calculate the brightness or intensity I_ν of a source, given its optical depth.[2]

---

[2]Equation (2.18) only describes the brightness emitted by a volume with a certain emissivity, given by Kirchhoff's law. In general, one would also have to account for the attenuation by any material located in the foreground.

$$I_\nu = B_\nu(T_e) \cdot (1 - e^{-\tau_\nu}). \tag{2.18}$$

In order to work out the spectral behaviour of the radiation (important to judge its contribution to the total radio emission), we consider the two extreme cases, namely the *optically thick* and the *thin* one. In the first case, we have $\tau_\nu \gg 1$ so that

$$I_\nu = B_\nu(T_e) = \frac{2\,h\,\nu^2}{c^2} \cdot k\,T_e.$$

The resulting emission spectrum follows exactly Planck's law in the Rayleigh-Jeans limit, i.e. we have a power-law as a function of frequency with slope $+2$. This spectrum is "featureless" and only depends on the electron temperature, with no other properties present! In the second case, we have $\tau_\nu \ll 1$ and

$$I_\nu = \tau_\nu \cdot B_\nu(T_e) = 8.29 \cdot 10^{-23} \cdot \left(\frac{EM}{\mathrm{pc\,cm^{-6}}}\right) \cdot \left(\frac{T_e}{10^4\,\mathrm{K}}\right)^{-0.5} \cdot g_{f\!f}.$$

This produces a radiation spectrum that varies only very slowly with frequency, dropping with increasing frequency $I_\nu \sim \nu^{-0.1}$, owing to the frequency dependence of the Gaunt factor. The above intensities or brightnesses have units of $\mathrm{erg\,s^{-1}\,cm^{-2}\,Hz^{-1}\,sr^{-1}}$. In radio astronomy, the term *"brightness temperature"* is commonly used. It is defined via the Rayleigh-Jeans approximation, and the temperature is therefore often referred to as the brightness temperature T_b.

$$I_\nu := \frac{2\,k\,\nu^2}{c^2} \cdot T_b \qquad \Longleftrightarrow \qquad T_b = \frac{c^2}{2\,k \cdot \nu^2} \cdot I_\nu.$$

We can now simply convert brightness into brightness temperature for the thermal free-free emission, using the above Rayleigh-Jeans approximation so that

$$\tau_\nu \gg 1 : I_\nu = B_\nu(T_e) \qquad \Longrightarrow \qquad T_b = T_e$$

$$\tau_\nu \ll 1 : I_\nu = \tau_\nu \cdot B_\nu \qquad \Longrightarrow \qquad T_b = \tau_\nu \cdot T_e.$$

We finally arrive at an important result, namely for the frequency dependence of the brightness and the brightness temperature (see Fig. 2.9). At low radio frequencies the optical depth is large, the intensity exhibits the slope of the Rayleigh-Jeans law, ν^2, and the observed brightness temperature equals the physical temperature. Above the turnover frequency, the brightness drops slowly as a function of frequency, $\nu^{-0.1}$, and the brightness temperature drops off as $\nu^{-2.1}$. The conversion of the so-called antenna temperature (the quantity that radio astronomers actually measure) into brightness temperature can be found in textbooks on radio astronomy.

We now know the frequency dependence of the thermal free-free radiation, which as we shall see is fundamentally different from that of the synchrotron radiation. This difference allows the two components, which in the Milky Way

Fig. 2.9 Continuum spectra
of free-free radiation

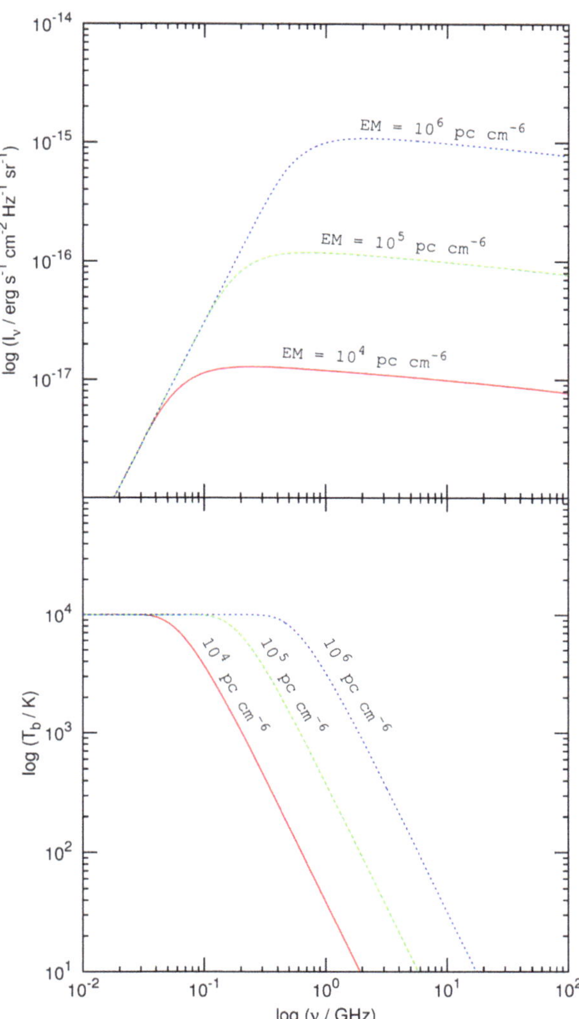

and in galaxies are always superimposed, to be separated. Even though this
separation is difficult (see the radio continuum spectrum of a galaxy in Fig. 2.31)
and requires multi-frequency measurements with very good signal-to-noise ratios,
it can be accomplished, given the data quality that is meanwhile achieved. The
other difference between the two radiation components is that free-free radiation is
unpolarised (owing to the random motions of the free electrons), while synchrotron
radiation is *linearly polarised* (see Sect. 2.3). So for free-free radiation we keep in
mind:

$$\left.\begin{array}{l} I_\nu \sim \nu^2 \\ T_b = T_e \end{array}\right\} \qquad \text{for} \qquad \tau_\nu \gg 1$$

$$\left.\begin{array}{l} I_\nu \sim \nu^{-0.1} \\ T_b \sim \nu^{-2.1} \end{array}\right\} \qquad \text{for} \qquad \tau_\nu \ll 1.$$

Measuring the brightness I_ν or the brightness temperature T_b over a sufficient frequency range allows to determine the emission measure, which is the dominant factor in the optical depth, which depends on the physical parameter $\langle n_e^2 \rangle^{1/2}$, i.e. the variance of the number density of the thermal electrons in an astrophysical plasma. Note that the dependence of the emission measure on n_e^2 is in the end due to the collisions occurring in the plasma!

In Fig. 2.10 (top) the thermal free-free radiation from the Large Magellanic Cloud (measured at a high radio frequency at which the synchrotron radiation is almost negligible in this galaxy) is juxtaposed to its Hα emission. The HII regions clearly show up in both images and are traced nearly identically. In Fig. 2.10 (middle) the thermal radiation emitted by the central region of the Orion Nebula is shown in the optical where the Balmer-line emission is preponderant, and in the radio regime showing the free-free continuum radiation. The radio continuum spectrum clearly shows the turnover frequency separating the optically thick from the optically thin frequency range. Figure 2.10 (bottom) shows the imprint of thermal absorption in the starburst galaxy M 82, seen as a 'cavity' in the brightness distribution of the nonthermal synchrotron radiation at 408 MHz. This ionised bubble marks the footpoint of the outflow of hot gas seen in the X-ray emission of this galaxy.

2.3 Synchrotron Radiation

As stated at the beginning of this chapter, synchrotron radiation is the main diagnostic tool to trace magnetic fields in the ISM and IGM. It is produced by the relativistic electrons that have been energised in the turbulent ISM and IGM via shock waves, notably caused by supernova explosions in the former case, and by active galactic nuclei (AGN), by galactic wakes and by merging of galaxy clusters in the latter. While moving at relativistic speeds, the particles are subject to the Lorentz force, which forces them into helical paths. The corresponding acceleration makes them radiate synchrotron radiation, which has a characteristic frequency spectrum and is partially polarised. The power of this diagnostic tool to trace magnetic fields is manyfold: galaxies and AGN exhibit synchrotron radiation as soon as they come into existence, a mere few hundred million years after the Big Bang. In fact, the distribution of faint (hence distant) radio sources is characterised by near isotropy, in accord with the cosmological principle. The bulk of these sources are AGN, which may produce Doppler boosting (see Sect. 7.2), which renders them detectable out to cosmological distances. Another mechanism that makes synchrotron radiation

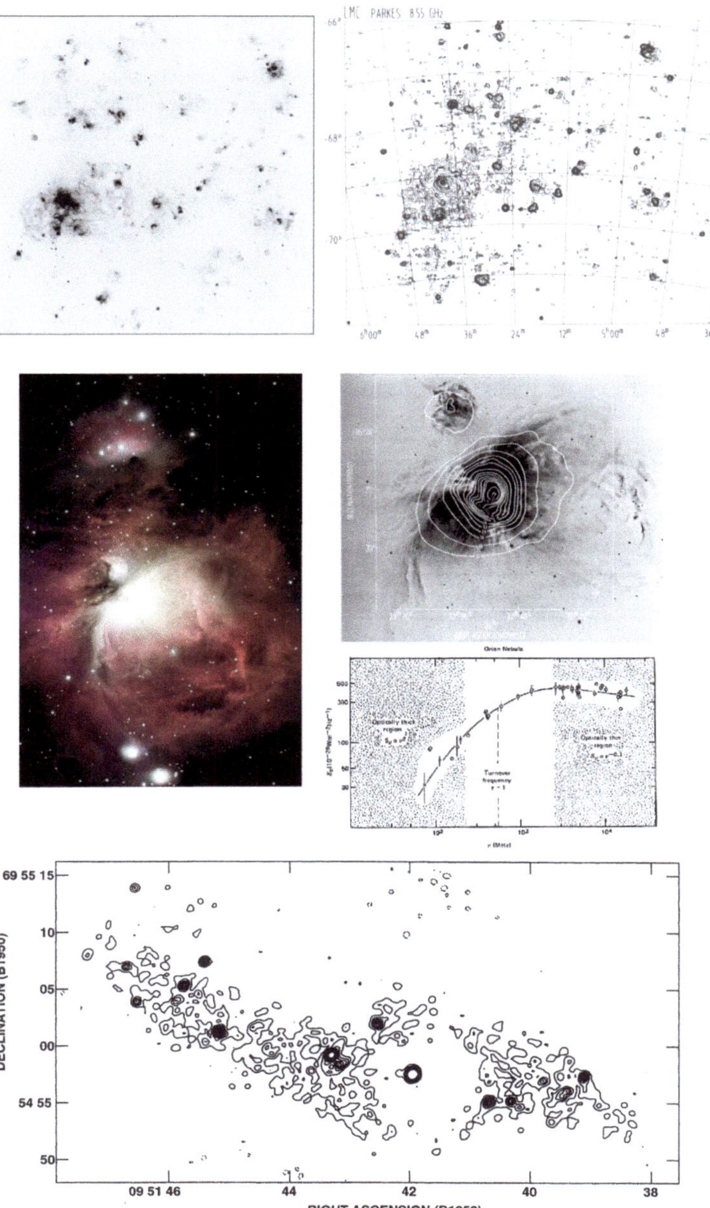

Fig. 2.10 *Upper row*: The Large Magellanic Cloud in Hα (*left*, from Kim et al. 1998), and at 8.6 GHz (*right*, from Haynes et al. 1991). *Middle row*: Centre of the Orion Nebula in Hα (*left*), contours of thermal free-free radio emission at 23 GHz, superimposed onto an optical image (*right*, from Wilson and Pauls, 1984), with the radio continuum spectrum below it, showing the low-frequency turnover due to free-free absorption. *Bottom*: Imprint of thermal absorption in M 82 (from Wills et al. 1997), seen as a 'cavity' in the overall nonthermal synchrotron radiation at 408 MHz

a very powerful tool to trace magnetism in diffuse media is Faraday rotation (Sect. 3.3), which allows us to estimate magnetic-field strengths and determine their orientation (towards or away from us) in the medium towards the radio source.

In what follows, we shall work out these important properties of synchrotron radiation on the basis of electro-magnetic theory.

2.3.1 Radiation from a Single Electron

As in the preceding section, we start out from the radiation by a single particle, this time accounting for its relativistic motion and energy. The power radiated into a unit solid angle per unit frequency and unit time is, as we have seen from (2.3),

$$\frac{dP}{d\Omega} = \frac{e^2}{4\pi c} \cdot \frac{|\vec{n} \times [(\vec{n} - \vec{\beta}) \times \dot{\vec{\beta}}]|^2}{(1 - \vec{n} \cdot \vec{\beta})^5}.$$

When dealing with relativistic particles, we have to distinguish the linear accelerator, in which case $\dot{\vec{\beta}} \parallel \vec{\beta}$ and the acceleration is caused by an electric field, and the transverse accelerator, in which case we have $\dot{\vec{\beta}} \perp \vec{\beta}$ and the acceleration is due to a magnetic field. The former is used in laboratory accelerators such as *Tevatron* or the planned *ILC*, while the latter is found in the form of cyclotrons and synchrotrons, e.g. the *LHC*, and in the ISM and IGM. We briefly consider the *linear accelerator*, which accelerates the particles in an electric field. In this case, we have

$$\frac{dP}{d\Omega} = \frac{e^2 \dot{v}^2}{4\pi c^3} \cdot \frac{\sin\theta}{(1 - \beta\cos\theta)^5},$$

and the radiation pattern has a strong dependence on the angle θ and on the particle speed. Figure 2.11 shows its three-dimensional illustration for $\beta = 0.8$, corresponding to $\gamma = 1.67$. This is mildly relativistic. In case of $\beta = 0.999$ ($\gamma = 22.4$), the radiation pattern would stretch out to a stunning 47,000 km on the scale of Fig. 2.11, more than the circumference of the earth! The reason is the very strong dependence of the maximum power radiated into the forward direction,

$$\frac{dP}{d\Omega}(\theta_{max}) \sim \gamma^8,$$

where

$$\cos\theta_{max} = \frac{1}{3\beta} \cdot \left(\sqrt{1 + 15\beta^2} - 1\right)$$

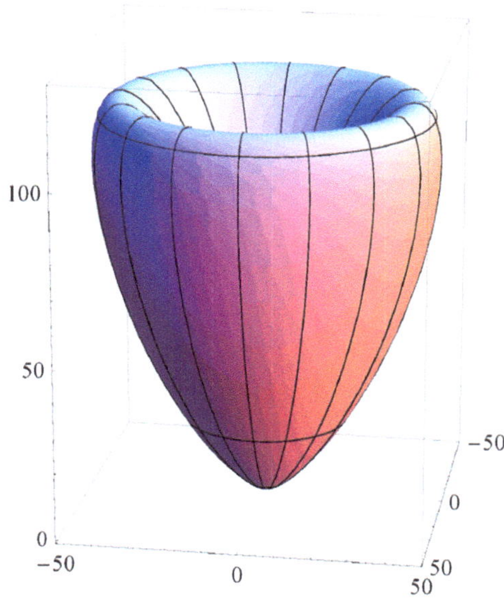

Fig. 2.11 Radiation pattern of the linearly accelerated electron ($\beta = 0.8$)

and

$$\theta_{max} = \frac{1}{2\gamma}.$$

The strong dependence on the Lorentz factor is called 'relativistic boosting' or 'beaming', meaning that a charged particle moving with relativistic speed emits essentially its whole radiation in the forward direction. In order to obtain the total radiation of the relativistic particle, we integrate over the solid angle again and obtain

$$P(t) = \int\limits_{0}^{2\pi} \int\limits_{0}^{\pi} \frac{dP}{d\Omega}\, d\Omega = \frac{2}{3} \cdot \frac{e^2 \dot{v}^2}{c^3} \cdot \gamma^6,$$

where

$$\gamma = \frac{1}{\sqrt{1-\beta^2}}.$$

is the Lorentz factor. We realise the cardinal difference to the non-relativistic case (2.6). In both cases, the total radiated power of a single particle depends on the square of the acceleration, $\dot{v}^2$. In addition, there is the very strong dependence on the Lorentz factor γ.

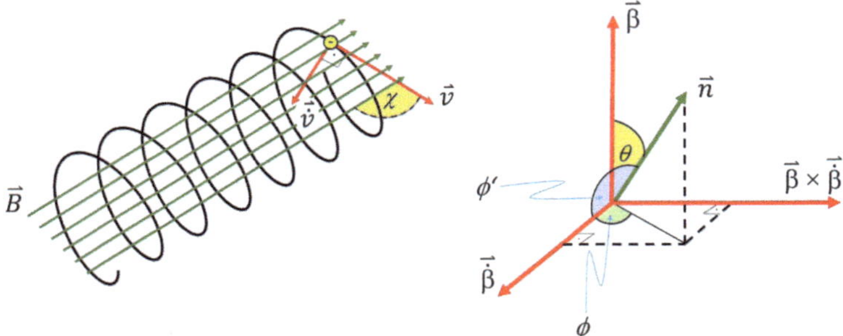

Fig. 2.12 Illustration of the various angles used in describing the transverse acceleration of a relativistic electron in a magnetic field

We now turn to the *transverse accelerator*, which is the more important process in the context of this book. In Fig. 2.12 the geometry of a relativistic particle moving in a magnetic field is illustrated. Here, $\theta = \angle(\vec{n}, \vec{\beta})$, $\phi' = \angle(\vec{n}, \dot{\vec{\beta}})$, ϕ is the angle measured along a circle perpendicular to $\vec{v}$, and θ is the angle measured in the $(\vec{v}, \dot{\vec{v}})$-plane. We are considering particles moving in the interstellar magnetic field, the trajectory inclined by angle χ with respect to the magnetic field vector. We refer to this angle as the 'pitch angle'. The charged particles are subject to the Lorentz force. This implies that $\vec{\beta} \perp \dot{\vec{\beta}}$ and (2.3) becomes

$$\frac{dP}{d\Omega} = \frac{e^2\,\dot{v}^2}{4\,\pi\,c^3}\cdot\frac{1 - \frac{\sin^2\theta\,\cos^2\phi}{\gamma^2\,(1-\beta\cos\theta)^2}}{(1-\beta\cos\theta)^3}.$$

As in case of the linear accelerator, the relativistic motion causes a relativistic aberration of the radiation of the charged particle, i.e. a strong distortion of the radiation pattern. In case of the transverse accelerator, this relativistic beaming produces a radiation pattern as illustrated in Fig. 2.13. Its main lobe can be shown to have a half-power width[3] that is inversely proportional to the Lorentz factor $1/\gamma$ at half-maximum:

$$\theta_{HP} \approx \frac{1}{\gamma} = \frac{m_0\,c^2}{E^2}. \qquad (2.19)$$

For example, consider a particle with energy $E = 1\,\text{GeV}$, which as we shall see below is the typical energy of relativistic electrons in the ISM producing synchrotron radiation at GHz frequencies. Such a particle has $\gamma = 2000$ ($\beta = 0.99999975$).

---

[3]The half-power width is the angular width of the radiation pattern at which the power has dropped to half its maximum value.

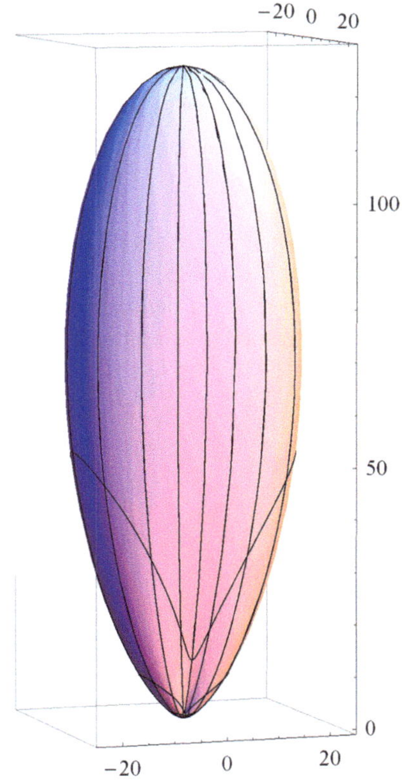

Fig. 2.13 Radiation pattern of the transversely accelerated electron ($\beta = 0.8$)

Hence, the opening angle of its radiation pattern has $\theta_{HP} = 1\overset{\circ}{.}7$. As in case of the transverse accelerator, the radiated power has a strong dependence on the Lorentz factor:

$$\frac{dP}{d\Omega} \sim \dot{v}^2 \gamma^6,$$

and

$$P(t) = \int\limits_0^{2\pi} \int\limits_0^{\pi} \frac{dP}{d\Omega} \cdot d\Omega \sim \dot{v}^2 \gamma^4. \tag{2.20}$$

A particle with $\gamma = 2000$ has a radiation pattern the maximum of which stretches out to some 8.6 light years on the scale of Fig. 2.13!

Apart from relativistic aberration, we have to apply the laws of special relativity in the transition from the reference frame of the particle to that of the observer, which turns out to significantly change the time dependence of the radiation seen in the observer's frame of reference. Before we look at that, we first have to work out the so-called Larmor circle, i.e. the circle on which the charged particle is moving about the magnetic field (see Fig. 2.14, left). We start out from the equation of motion, given by the Lorentz force:

$$m\,\dot{\vec{v}} = m \cdot (\vec{v} \times \vec{\omega}_L) = -\frac{e}{c} \cdot (\vec{v} \times \vec{B}).$$

Here, m is the mass of the particle (electron in our case), ω_L is the Larmor frequency, $\vec{v}$ its velocity and $\vec{B}$ the vector of its magnetic field strength. In case of $\chi = 90°$ i.e. a particle motion perpendicular to the magnetic field the above equation can be rewritten in the form

$$m\,\omega_L^2\,r_L = m \cdot \frac{v^2}{r_L} = \frac{e}{c} \cdot v\,B$$

$$m\,\frac{v}{r_L} = \frac{e\,B}{c}$$

$$\omega_L = \frac{e\,B}{m\,c}.$$

Note that since the particle is relativistic we have $m = m_0\,\gamma$, since $E = m\,c^2 = \gamma\,m_0\,c^2$ (m_0 is the rest mass of the electron). It is then obvious that the Larmor frequency decreases with increasing Lorentz factor γ:

$$\omega_L = \frac{e\,B}{\gamma\,m_0\,c}.$$

From the above, the Larmor radius is obtained via ($v \approx c$)

$$r_L = \frac{v}{\omega_L} = \frac{m_0\,v\,c}{eB} \cdot \gamma \approx \frac{m_0\,c^2}{e\,B} \cdot \gamma = \frac{E}{e\,B},$$

or, in general ($\chi \neq 90°$)

$$\boxed{r_L = \frac{E}{e\,B} \cdot \sin \chi.} \qquad (2.21)$$

In Table 2.1 the Larmor radii and frequencies are listed for different energies for a magnetic-field strength of $B = 10\,\mu\text{G}$.

We realise that the Larmor radius does not depend on the mass of the particle, but just on its energy (and on the magnetic-field strength). The particles moving on their helical orbits about the interstellar magnetic field produce radiation that depends on the temporal behaviour given by Eq. (2.20), with the radiation pattern shown in Fig. 2.13 sweeping around with an angular speed corresponding to the Larmor frequency. This frequency corresponds to a mere sweeping frequency of

Table 2.1 Larmor radii and frequencies ($B = 10\,\mu$G)

E (eV)	r_L	ν_L (Hz)	γ
10^9	$3.3 \cdot 10^6$ km	$1.4 \cdot 10^{-2}$	2000
4.5^{10}	$1.5 \cdot 10^8$ km	$3.1 \cdot 10^{-4}$	$9 \cdot 10^4$
10^{20}	10 kpc	$1.4 \cdot 10^{-13}$	$2 \cdot 10^{14}$

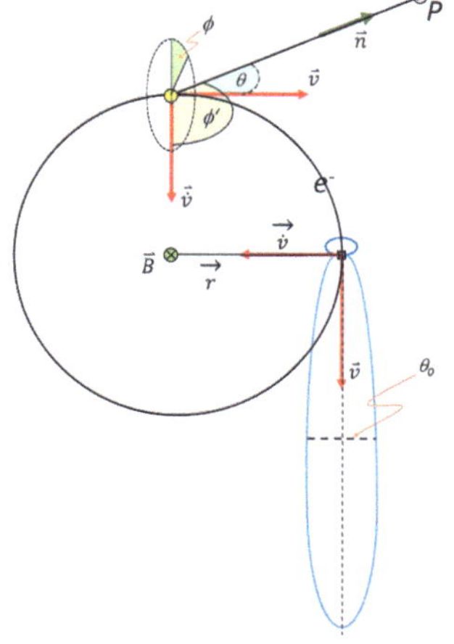

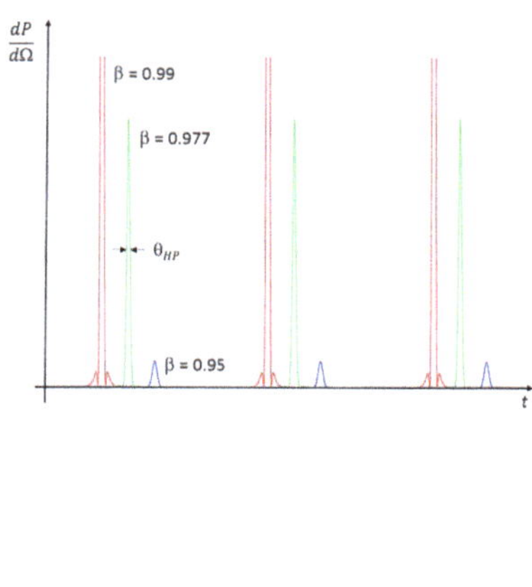

Fig. 2.14 *Left*: Geometry of the Larmor circle. *Right*: Sketch of the emitted pulses of the gyrating relativistic electrons as seen in the particles' reference frame. Pulse sequences are shown for three different values of β, and have been shifted for the sake of clarity

28 Hz for an electron with $\gamma = 1$ in a magnetic field with $B = 10\,\mu$G. The reader might now ask the question how, then, we can detect synchrotron radiation from such particles in the radio regime? In order to see this we need to calculate the time dependence of the radiation as seen by the observer.

In order to calculate the radiation spectrum, we have to follow the same path as in Sect. 2.2.2, i.e. we perform a Fourier analysis of the time-dependent radiation power of single particles, then 'fold in' their energy spectrum and then calculate the emissivity. The frequency of the emitted pulses of the gyrating relativistic electrons (Fig. 2.14, right) corresponds to the inverse of the time that the radiation pattern needs to sweep across the observer. Note in Fig. 2.14 how the amplitude is boosted by a slight increase of β. The spike for $\beta = 0.99$ is already way outside the frame of the graph. The duration of the pulse in the particle's frame of reference is equal to

$$\Delta t = \frac{r_L \, \theta_{HP}}{v} \approx \frac{r_L \, \theta_{HP}}{c}.$$

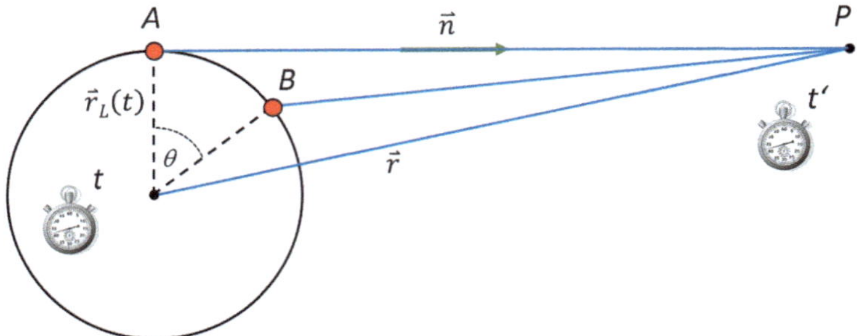

Fig. 2.15 Illustration of the geometry of the transformation from the particle's to the observer's reference frame

Since $r_L \approx E/e\,B$ and $\theta_{HP} \approx 1/\gamma$, we find

$$\Delta t = \frac{m_0\,c}{e\,B}.$$

For example, this is about $6\,\text{ms}$ in a $10\text{-}\mu\text{G}$ magnetic field, corresponding to $\nu \approx 180\,\text{Hz}$. We now transform this time duration from the particle's reference frame t to that of the observer t' (Fig. 2.15).

In order to do that we have to account for the motion of the particle while we measure the pulse width Δt, which means that t' is equal to t plus the time elapsed while the particle moves from A to B as shown in (Fig. 2.15), i.e.

$$t' = t + \frac{|\vec{r} - \vec{r}_L(t)|}{c}.$$

This enables us to establish the relation between the two frames[4]

$$\frac{dt'}{dt} = 1 - \frac{1}{c}\frac{\vec{r} - \vec{r}_L(t)}{|\vec{r} - \vec{r}_L(t)|} \cdot \frac{d\vec{r}_L(t)}{dt}$$

$$= 1 - \frac{\vec{n} \cdot \vec{v}}{c} = 1 - \beta \cdot \cos\theta_{HP} \approx 1 - \beta \cdot \sqrt{1 - \theta_{HP}^2}$$

$$= 1 - \beta \cdot \sqrt{1 - \frac{1}{\gamma^2}} = 1 - \beta^2 = \frac{1}{\gamma^2},$$

and we finally arrive at the important result that

---

[4] Here we make use of the derivative $\frac{d}{dt}\left\{[\vec{r} - \vec{r}_L(t)]^2\right\}^{\frac{1}{2}}$.

$$\Delta t' = \frac{\Delta t}{\gamma^2} \, .$$

We now see that the frequency spectrum is shifted towards a γ^2-times higher range, so that for example an electron with an energy of $E = 1\,\text{GeV}$, corresponding to $\gamma = 2000$, radiates at a frequency of $\nu = 700\,\text{MHz}$ in the presence of a 10-μG magnetic field! A more precise determination of the radiation spectrum (of a single electron) requires a Fourier analysis (such as in the non-relativistic case).

We define the critical frequency such that the particles produce significant power at that frequency $\omega_c = 2\,\pi\,\nu_c$. There are different definitions around in the literature, which are due to three renowned physicists who worked on the theory of synchrotron radiation, viz. J. Schwinger, V. L. Ginzburg and J. D. Jackson:

$$\omega_c \overset{\text{Schwinger}}{:=} \frac{1}{\frac{2}{3} \cdot \Delta t'}$$

$$\omega_c \overset{\text{Ginzburg}}{:=} \frac{1}{\Delta t'}$$

$$\omega_c \overset{\text{Jackson}}{:=} \frac{1}{\frac{1}{3} \cdot \Delta'} \, .$$

In what follows, we will use Schwinger's (1949) definition. With (2.22) we then derive a critical frequency that depends on the strength of the magnetic field and on the Lorentz factor:

$$\boxed{\nu_c = \frac{3}{4\,\pi} \cdot \frac{e\,B_\perp}{m_0\,c} \cdot \gamma^2. \qquad (2.22)}$$

Here

$$B_\perp = B \cdot \sin\chi$$

is the component of the magnetic field perpendicular to the line-of-sight. In Table 2.2 a few values for critical frequencies are compiled, assuming a magnetic-field strength of $B = 10\,\mu$G, which is a typical value for the ISM. In Table 2.2, critical frequencies are listed for a range of particle energies.

Table 2.2 Particle energies and their critical frequencies

E (GeV)	γ	ν_c
1	200	170 MHz
45	$9 \cdot 10^4$	340 GHz
10^{11}	$2 \cdot 10^{14}$	$1.7 \cdot 10^{30}$ Hz

So far, the calculations towards the radiation spectrum of relativistic particles in the ISM have been performed assuming electrons only. What about protons (or other nuclei)? The cosmic-ray (CR) energy spectrum observed near earth exhibits ~ 100 times more protons than electrons (at the same energy). Nevertheless, they do not contribute significantly to the emission, since

$$\nu_c = \frac{3}{4\pi} \cdot \frac{e\,B_\perp}{m_0\,c} \cdot \gamma^2 = \frac{3}{4\pi} \cdot \frac{e\,B_\perp E^2}{m_0^3\,c^5}.$$

Thus, ν_c depends on the mass of the radiating particle like m^{-3}:

$$\left(\frac{m_p}{m_e}\right)^{-3} = 1.6 \cdot 10^{-10},$$

hence

$$\frac{\nu_{c,p}}{\nu_{c,e^-}} = 1.6 \cdot 10^{-10}.$$

Put differently, we can calculate how much more kinetic energy a proton must have in order to radiate at the same frequency as the electron.

$$E_p = \left(\frac{m_p}{m_e}\right)^{3/2} \cdot E_e = 8 \cdot 10^4 \cdot E_e.$$

Hence, even the ratio of number densities measured in the CR energy spectrum of $n_p/n_e \approx 100$ does not help. In fact, as we shall see later, relativistic protons are much more long-lived, owing to their very low radiation losses. They may remain relativistic for more than a Hubble time, while electrons become non-relativistic within less than 100 Myr.

The Fourier analysis of the time-dependent radiated power (2.3) is obtained by using its approximation for small angles θ and large Lorentz factors γ, which is

$$\frac{dP}{d\Omega} = \frac{2}{\pi}\frac{e^2\,\dot{v}^2}{c^3}\gamma^6 \cdot \frac{1}{(1+\gamma^2\,\theta^2)^3} \cdot \left[1 - \frac{4\,\gamma^2\,\theta^2\,\cos^2\phi}{(1+\gamma^2\,\theta^2)^2}\right],$$

and upon integration over solid angle becomes

$$P(t) = \int_0^{2\pi}\int_0^{\pi} \frac{dP}{d\Omega}\,d\Omega = \frac{2}{3} \cdot \frac{e^2\,\dot{v}^2}{c^3} \cdot \gamma^4.$$

The tedious Fourier analysis of this expression has been done for us by J. Schwinger, yielding

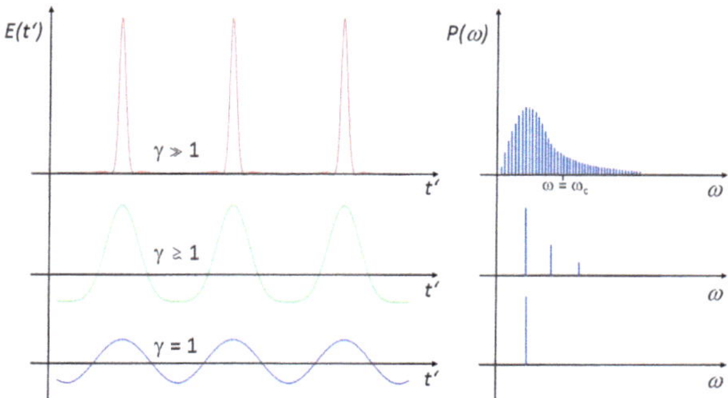

Fig. 2.16 Sketch of the time dependence of the synchrotron pulses and their radiation spectra

$$P(\nu) = \frac{\sqrt{3}\,e^3}{m_0\,c^2} \cdot B_\perp \cdot F\left(\frac{\nu}{\nu_c}\right),$$

where

$$F\left(\frac{\nu}{\nu_c}\right) = \frac{\nu}{\nu_c} \cdot \int\limits_{\nu/\nu_c}^{\infty} K_{5/3}(x)\,dx.$$

The function $F(\frac{\nu}{\nu_c})$ is the so-called Airy integral of the modified Bessel function $K_{5/3}(x)$. It is well approximated by the so-called Wallis approximation (Wallis, 1959), obtained after some lengthy algebra:

$$F\left(\frac{\nu}{\nu_c}\right) = 1.78 \left(\frac{\nu}{\nu_c}\right)^{0.3} \cdot e^{-\frac{\nu}{\nu_c}}. \tag{2.23}$$

In Fig. 2.16, some synchrotron pulses emitted by relativistic particles with increasing Lorentz factor are sketched, with the corresponding radiation spectra juxtaposed. In case of $\gamma = 1$, trivially, the frequency spectrum consists of a single component (δ-function) only, since the 'pulses' are simply sinusoidal. As γ increases, more and more frequency components are added, owing to the increasing distortion of the pulses. For $\gamma \gg 1$ we obtain a quasi-continuum of frequencies, its envelope given by (2.23).

2.3.2   *Synchrotron Radiation from Relativistic Electrons with an Energy Spectrum*

In order to calculate the radiation spectrum of an ensemble of relativistic electrons, we need to know their energy spectrum. This has been measured in the earth's

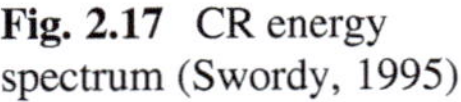

Fig. 2.17 CR energy spectrum (Swordy, 1995)

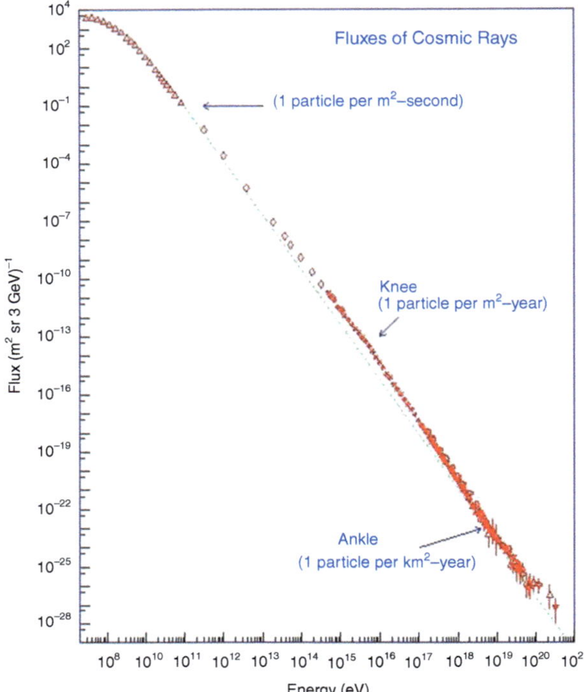

vicinity by studying the CRs raining down onto the atmosphere. From such measurements, a power-law of the following form has been found:

$$N(E)\,dE = A \cdot E^{-g}\,dE. \qquad (2.24)$$

Here, A is a 'constant'.[5] It contains the local number density of relativistic particles per energy interval. The power-law index is generally found to be $g = 2.4$, which will turn out to be very convenient in our computation below. In Fig. 2.17 the measured energy spectrum near earth is displayed. The energies of particles producing the galactic synchrotron radiation are in the range of $\leq 1\,\mathrm{GeV}$ to several tens of GeV. Unfortunately, the measured spectrum is strongly modulated by the solar wind below a few GeV, which explains the deviation from the power-law there. Hence, nothing is known about the shape of the spectrum at the lowest CR energies. At the highest energies, there are changes in the spectrum called 'knee' (at $\gtrsim 10^{15}\,\mathrm{eV}$) and 'ankle' (at $\gtrsim 10^{18}\,\mathrm{eV}$). The particles with the highest recorded energies ($\gtrsim 10^{20}\,\mathrm{eV}$, so-called ultra-high energy cosmic rays, or UHECR) are a real enigma, their origin being totally unknown.

---

[5]The term 'constant' refers to the local conditions near earth.

Next, we calculate the intensity via the emissivity as follows:

$$4\,\pi\,\epsilon_\nu = \int_{E_1}^{E_2} P(\nu) \cdot N(E)\,dE.$$

Assuming that there is no background radiation, the radiation transport equation yields the intensity from the brightness and the source function, the latter given by the emission and absorption coefficient.

$$I_\nu = S_\nu(T) \cdot (1 - e^{-\tau_\nu}) \approx S_\nu(T) \cdot \tau_\nu.$$

With Kirchhoff's law

$$S_\nu(T) = \frac{\epsilon_\nu}{\varkappa_\nu}$$

we have

$$I_\nu = \int_0^{s_0} \epsilon_\nu\,ds,$$

and hence

$$I_\nu = \frac{1}{4\,\pi} \int_0^{s_0} \int_0^{\infty} P(\nu)\,N(E)\,dE\,ds. \tag{2.25}$$

It is readily seen that this brightness or intensity has dimension $\mathrm{erg\,s^{-1}\,cm^{-2}\,Hz^{-1}\,sr^{-1}}$. Let us assume for simplicity that neither the power $P(\nu)$ nor the energy spectrum $N(E)$ depend on location, i.e. they are independent of s, or $dP/ds = dN/ds = 0$. Then the above integral reduces to

$$I_\nu = \frac{s_0}{4\,\pi} \cdot \frac{\sqrt{3}\,e^3}{m_0\,c^2} \cdot B_\perp\,A \cdot \int_0^{\infty} F\left(\frac{\nu}{\nu_c}\right) E^{-g}\,dE,$$

where s_0 is the total path length. Plugging in Wallis' approximation delivers

$$I_\nu = \frac{s_0}{4\,\pi} \cdot \frac{\sqrt{3}\,e^3}{m_0\,c^2} \cdot B_\perp\,A \cdot 1.78 \cdot \int_0^{\infty} \left(\frac{\nu}{\nu_c}\right)^{0.3} \cdot e^{-\frac{\nu}{\nu_c}} \cdot E^{-g}\,dE.$$

We next define some quantities that render the formulae easier to handle by lumping numerical factors and constants into them.

$$C := 1.78 \cdot \frac{\sqrt{3} \cdot e^3}{4\pi\, m_0 \cdot c^2} = 3.32 \cdot 10^{-23} \text{ esu}^3 \text{ erg}^{-1},$$

$$\nu_c = \frac{3}{4\pi} \cdot \frac{e\, B_\perp}{m_0^3\, c^5} \cdot E^2 := \eta\, B_\perp\, E^2,$$

$$\eta = 6.26 \cdot 10^{18} \text{ s}^4 \text{ g}^{-5/2} \text{ cm}^{-7/2}.$$

Using the substitution

$$\sqrt{\frac{\nu_c}{\nu}} =: x = \left(\frac{\eta \cdot B}{\nu}\right)^{\frac{1}{2}} \cdot E,$$

i.e.

$$dE = \left(\frac{\nu}{\eta\, B}\right)^{\frac{1}{2}} dx,$$

we obtain the following expression for the intensity:

$$I_\nu = s_0\, C\, A\, \eta^{\frac{g-1}{2}} B_\perp^{\frac{g+1}{2}} \nu^{-\frac{g-1}{2}} \cdot \int_0^\infty x^{-(g+0.6)}\, e^{-\frac{1}{x^2}}\, dx. \tag{2.26}$$

It conveniently turns out that in the Milky Way, but also in external galaxies $g = 2.4$ so that the integral has a trivial solution! With

$$\frac{1}{x^2} = u, \qquad \text{i.e.} \qquad -\frac{2}{x^3} \cdot dx = du,$$

it follows that

$$\int_0^\infty x^{-3} \cdot e^{-\frac{1}{x^2}}\, dx = \frac{1}{2} \int_0^\infty e^{-u}\, du = \frac{1}{2}.$$

Inserting $g = 2.4$ we arrive at

$$I_\nu = 2.4 \cdot 10^{-10} \left(\frac{s_0}{\text{cm}}\right) \left(\frac{A}{\text{erg}^{1.4}\, \text{cm}^{-3}}\right) \left(\frac{B_\perp}{\text{G}}\right)^{1.7} \left(\frac{\nu}{\text{Hz}}\right)^{-0.7},$$

which in this form has dimension $\text{erg}\,\text{s}^{-1}\,\text{cm}^{-2}\,\text{Hz}^{-1}\,\text{sr}^{-1}$. Close to the earth, the constant in (2.24) takes the value $A = 8.2 \cdot 10^{-17}\,\text{erg}^{1.4}\,\text{cm}^{-3}$. If this constant would hold over a line-of-sight of $10\,\mathrm{kpc}$, then with a magnetic-field strength of $B = 10\,\mu\mathrm{G}$ we would expect a synchrotron intensity of

$$I_\nu \approx 10^{-18} \ \mathrm{erg \ s^{-1} \ cm^{-2} \ Hz^{-1} \ sr^{-1}}$$

at an observing frequency of $\nu = 1\,\mathrm{GHz}$, which is close to what we measure towards the Galactic plane. In general, with

$$N(E)\,dE \sim E^{-g}\,dE$$

we arrive at

$$I_\nu \sim B_\perp^{1+\alpha} \cdot \nu^{-\alpha}, \qquad (2.27)$$

revealing that – apart from any spatial variations that we have neglected in the above derivation – the synchrotron intensity depends on the strength of the magnetic-field component perpendicular to the line-of-sight, and on the frequency, with a spectral index α that relates to the power-law index g of the energy spectrum via

$$\alpha = \frac{g-1}{2}. \qquad (2.28)$$

Plugging $g = 2.4$ into the above equation, we obtain $\alpha = 0.7$. This value is actually almost invariably found in the ISM! The synchrotron spectrum steepens in regions of lacking energy supply (see Sect. 2.3.4), while in the vicinity of star-forming regions in which stellar winds and supernovae cause turbulence and (re-)accelerate the particles the canonical power-law radiation spectrum with $\alpha = 0.7$ is found. Note that the factor of 1/2 arises from the fact that $E \sim \nu^{1/2}$.

What is happening in computing I_ν is that for each electron the radiation spectrum $P(\nu)$ of the single particle is successively multiplied by the particles' number density for each energy (see Fig. 2.18). The integration over the whole energy range then yields the frequency spectrum. In the log-log plot this means that we have to add (logarithmically) the 'weighting functions', given by $N(E)$. If the energy spectrum has a cut-off at some energy E_{max}, the spectrum will fall off exponentially beyond the corresponding critical frequency (Fig. 2.19),

$$\nu_c = \frac{3}{4\,\pi} \cdot \frac{e \cdot B_\perp}{m_0\,c} \cdot \gamma_{max}^2.$$

Examples of synchrotron radiation are shown in Figs. 2.20 and 2.21. The Aitoff projection of the radio emission from the Milky Way at $408\,\mathrm{MHz}$ is dominated by synchrotron radiation almost everywhere, except for the most intense ridge along the Galactic plane, where there is a significant contribution by thermal free-free radiation (see Chap. 4). A large number of discrete sources is seen superimposed onto the diffuse emission. Some of these are located in the Galaxy, while the bulk

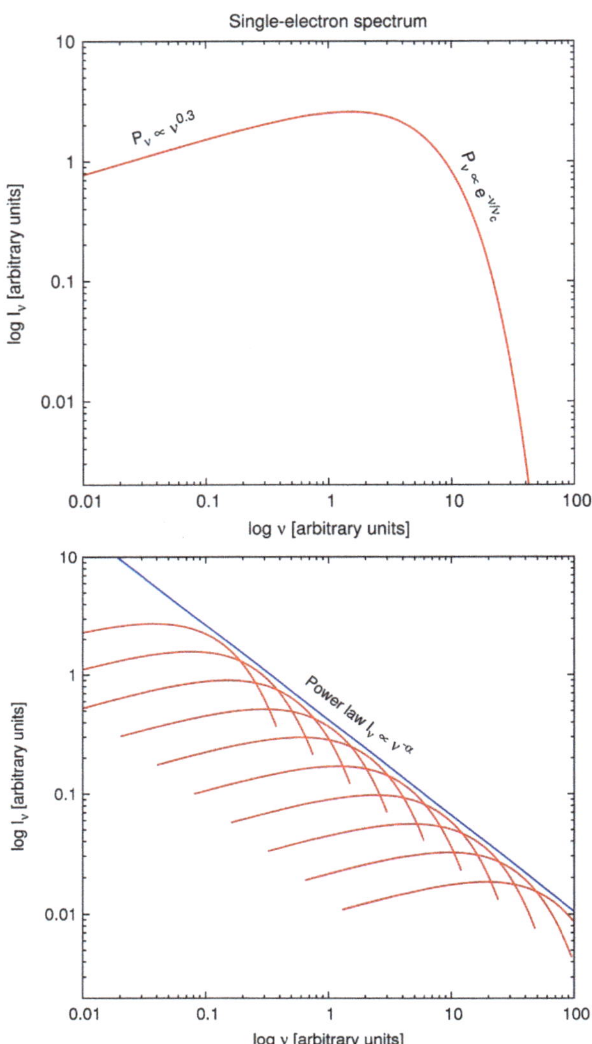

Fig. 2.18 Illustration of the radiation spectrum

of them is extragalactic. Fornax A is a classical radio galaxy with two extended 'radio lobes', which are powered by a central AGN that must be connected with a supermassive black hole (see Chap. 7).

2.3.3 Polarisation Properties

As we have seen in the previous section, relativistic particles radiate within a narrow beam of width $\theta_{HP} = \gamma^{-1}$. Considering their helical motion around the magnetic

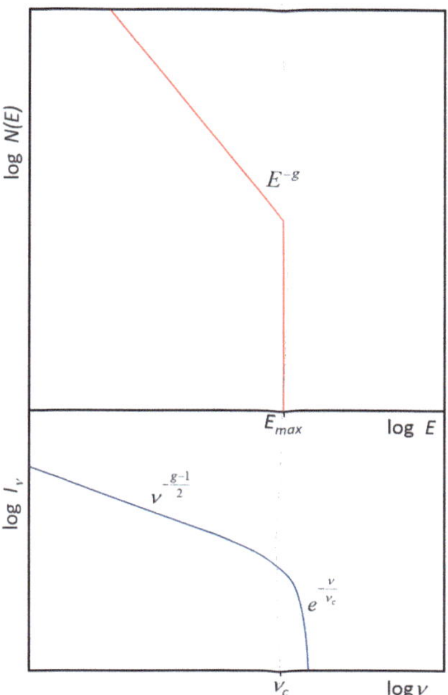

Fig. 2.19 Radiation spectrum with energy cutoff

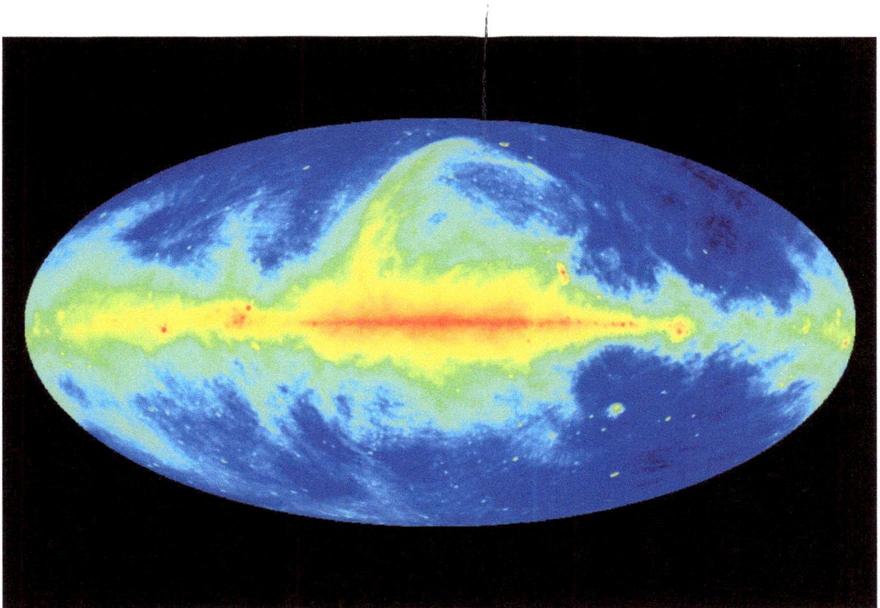

Fig. 2.20 Radio emission from the Milky Way at 408 MHz (Haslam et al. 1982)

Fig. 2.21 Radio emission of the radio galaxy Fornax A (colour), superimposed onto an optical image (Fomalont et al. 1989)

field, the radiation is emitted into a velocity cone (Fig. 2.22), which is the cone described by the velocity vector $\vec{v}$. Its axis is along the direction of the magnetic-field vector. Hence, the opening angle of the cone is twice the pitch angle χ. The cone axis is parallel to the magnetic field, and $\vec{v}$ precesses about this direction with the gyro-frequency. In the non-relativistic case, looking along the magnetic field lines we would measure circular polarisation, or, if viewing at some angle w.r.t the magnetic field $\vec{B}$, elliptical polarisation. The fundamental difference in case of relativistic particles is that significant radiation is only measured if the trajectory of the electrons lies within the (very narrow) angle $1/\gamma$ of the line of sight. The geometric situation is illustrated in Fig. 2.23.

The details are rather complicated. The full algebra can be found in Longair (2011). The best way to understand how a single particle produces elliptical polarisation and how an ensemble of particles with a pitch-angle distribution

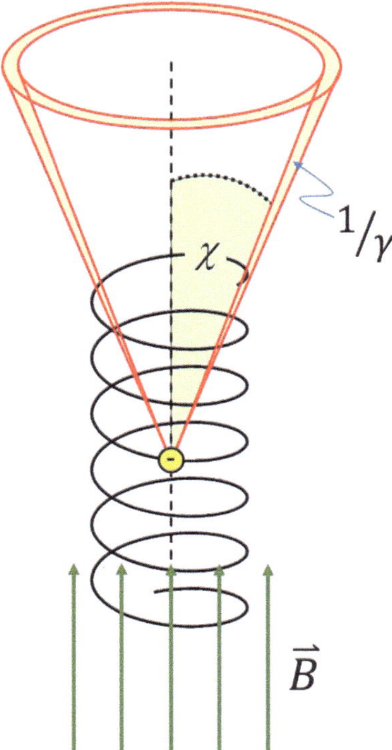

Fig. 2.22 Velocity cone of a particle moving in a magnetic field

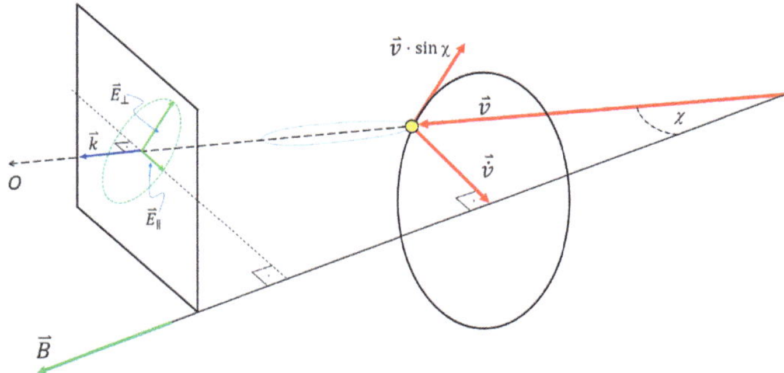

Fig. 2.23 Orientation of the electric field $\vec{E}$ produced by the particle gyrating in a magnetic field $\vec{B}$

produces net linear polarisation is by looking at three-dimensional velocity cones (paper models). In case of non- or mildly relativistic particles the radiation pattern would be broad and we would observe the (rotating) $\vec{E}$-vector over the full gyro-circle around the magnetic field. If the particles are highly relativistic, we see the light pulse of width $1/\gamma$ only for a *very* short time, hence we see it only from particles with *one specific pitch angle.* and thus record the pulse only over a very small fraction of the velocity cone.

In order to see how the particles produce linear polarisation, let us first consider the case of those electrons whose velocity cones lie precisely along the line-of-sight to the observer (Fig. 2.23). When the electron points directly to the observer, its acceleration $\dot{\vec{v}}$ is in the direction $\vec{v} \times \vec{B}$, so that the observed radiation is linearly polarised, the electric field is oriented along the direction $\vec{v} \times \vec{B}$, while it lies in the plane perpendicular to the wave vector $\vec{k}$. Hence, in this case, the $\vec{E}$-vector is perpendicular to the projection of $\vec{B}$ onto the plane of the sky.

If the electron does not precisely point to the observer within the velocity cone, then there is also a component of the electric field parallel to the magnetic-field direction. The radiation from a single electron is elliptically polarised because the component parallel to the field has a different time dependence within each pulse compared with that of the perpendicular component. This is reflected in the fact that the frequency spectra of the two polarisations are different (see below). When there is a distribution of pitch angles, however, all the electrons with velocity cones within an angle $1/\gamma$ around the line-of-sight contribute to the intensity measured by the observer. The total net polarisation is found by integrating over all particles which contribute to the intensity, If these particles are relativistic, we observe their radiation only for a very short time $\propto \gamma^{-2}$, during which their trajectory as seen by us is a straight line. Hence the resultant polarisation is linear.

The precise calculation of the degree of polarisation is again obtained via a Fourier analysis of $P_\perp(t)$ and $P_\parallel(t)$, yielding

$$P_\perp(v) = \frac{\sqrt{3}\, e^3\, B_\perp}{m_0\, c^2} \cdot [F(x) + G(x)]$$

$$P_\parallel(v) = \frac{\sqrt{3}\, e^3\, B_\perp}{m_0\, c^2} \cdot [F(x) - G(x)],$$

where

$$F(x) = x \cdot \int\limits_{x}^{\infty} K_{5/3}(z)\, dz$$

as before, and

$$G(x) = x \cdot K_{2/3}(x),$$

with $x = v/v_c{}^6$. In Fig. 2.24 a sketch of the functions $F(x)$ and $G(x)$ is shown. Their behaviour can be qualitatively understood as follows: since the amplitude of

---

[6]Note that when doing the integration to obtain the total (unpolarised) intensity (2.26), the definition was $x = (v/v_c)^{1/2}$.

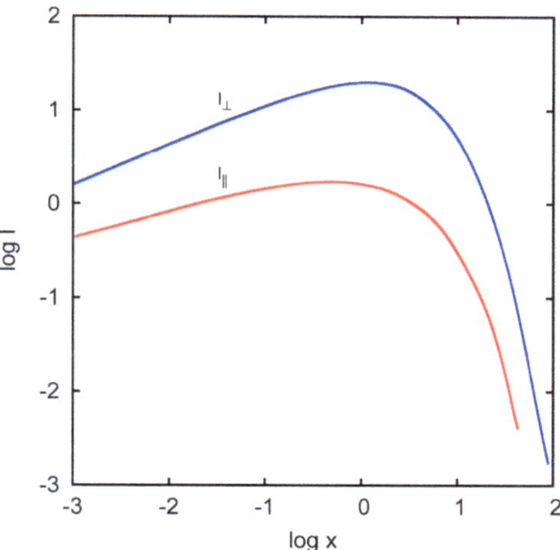

Fig. 2.24 Intensities radiated parallel and perpendicular to the magnetic field

the electric field component $E_\parallel$ is always smaller than that of the component $E_\perp$, the power $P_\parallel$ must be lower than that of $P_\perp$. Furthermore, since $E_\parallel$ varies more slowly as a function of time, the power $P_\parallel$ peaks at a lower frequency. The degree of linear polarisation at a single energy now results in

$$P(\nu) = \frac{p_\perp(\nu) - p_\parallel(\nu)}{p_\perp(\nu) + p_\parallel(\nu)}.$$

Integration over the power-law energy spectrum, i.e. $N(E)\,dE \sim E^{-g}\,dE$, yields

$$p = \frac{g + 1}{g + \frac{7}{3}} = \frac{\alpha + 1}{\alpha + \frac{5}{3}}. \qquad (2.29)$$

With $g = 2.4$, i.e. $\alpha = 0.7$, we expect a maximum degree of linear polarisation of a synchrotron source of $p = 72\%$. This assumes that the magnetic field is absolutely uniform and that there are no depolarising effects (Sect. 3.3.2) such as beam or Faraday depolarisation.

The polarisation properties of an electromagnetic wave are described by an ellipse swept out by the Cartesian components of the electric-field vector. The geometry is shown in Fig. 2.25. The orthogonal components of the electric field can be represented by

$$E_x = E_{x_0} \cdot \cos(k\,z - \omega t - \delta_1)$$
$$E_y = E_{y_0} \cdot \cos(k\,z - \omega t - \delta_2),$$

where k is the wave number, δ_1 and δ_2 the phases and z the direction of propagation of the wave. Writing $\tau = kz - \omega t$, we have

Fig. 2.25 Polarisation ellipse

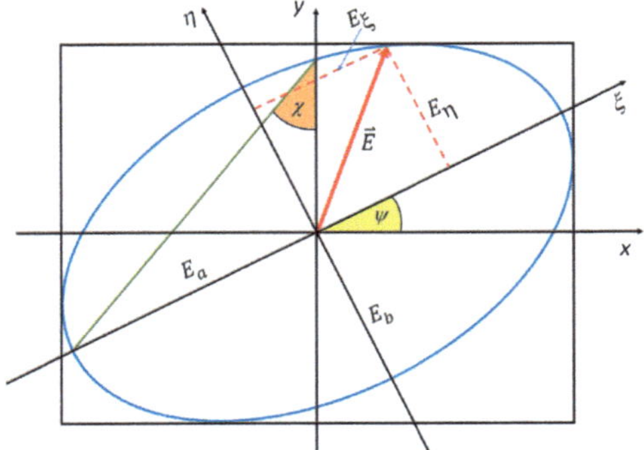

$$E_x = E_{x_0} \cdot \cos(\tau - \delta_1)$$

$$E_y = E_{y_0} \cdot \cos(\tau - \delta_2),$$

which translates into

$$\frac{E_x}{E_{x_0}} = \cos\tau \cdot \cos\delta_1 - \sin\tau \cdot \sin\delta_1 \tag{2.30}$$

$$\frac{E_y}{E_{y_0}} = \cos\tau \cdot \cos\delta_2 - \sin\tau \cdot \sin\delta_2. \tag{2.31}$$

We now multiply the first of the above equations by $\sin\delta_2$ and the second one by $\sin\delta_1$ and subtract these from one another; we then multiply the first of the above equations by $\cos\delta_2$ and the second one by $\cos\delta_1$ and also subtract these from one another. This leads to the following expressions:

$$\frac{E_x}{E_{x_0}} \cdot \sin\delta_2 - \frac{E_y}{E_{y_0}} \cdot \sin\delta_1 = \cos\tau \cdot \sin(\delta_2 - \delta_1) \tag{2.32}$$

$$\frac{E_x}{E_{x_0}} \cdot \cos\delta_2 - \frac{E_y}{E_{y_0}} \cdot \cos\delta_1 = \sin\tau \cdot \sin(\delta_2 - \delta_1). \tag{2.33}$$

Squaring and adding the above two equations, we obtain an *ellipse equation* of the form

$$\left(\frac{E_x}{E_{x_0}}\right)^2 + \left(\frac{E_y}{E_{y_0}}\right)^2 - 2\,\frac{E_x\,E_y}{E_{x_0}\,E_{y_0}}\cos\delta = \sin^2\delta,$$

where $\delta = \delta_2 - \delta_1$. Its orientation and shape are defined by the angles χ and ψ, respectively, where the latter is the polarisation angle and the former tells us how much circular polarisation is contained in the electro-magnetic wave. The electric field can be also expressed in terms of circularly polarised waves:

$$E_l = E_{l_0} \cdot \cos(k\,z - \omega\,t + \delta')$$

$$E_r = E_{r_0} \cdot \cos(k\,z - \omega\,t).$$

The Poynting flux is given by

$$S_0 = |\vec{E} \times \vec{H}|$$
$$= E_{x_0}^2 + E_{y_0}^2$$
$$= E_{r_0}^2 + E_{l_0}^2$$
$$= E_a^2 + E_b^2.$$

At this point it is convenient to introduce the Stokes parameters and the Poincaré sphere (Fig. 2.26). These definitions allow to measure the polarisation properties in a very simple way, by adding, subtracting, or multiplying the orthogonal or circular components of the electric field ($I = S_0$):

$$I = \langle E_a^2 \rangle + \langle E_b^2 \rangle \quad = \langle E_{x_0}^2 \rangle + \langle E_{y_0}^2 \rangle \quad = \langle E_{r_0}^2 \rangle + \langle E_{l_0}^2 \rangle \tag{2.34}$$

$$Q = I\,\cos 2\chi\,\cos 2\psi = \langle E_{x_0}^2 \rangle - \langle E_{y_0}^2 \rangle \quad = 2\,\langle E_{l_0}\,E_{r_0} \rangle\,\cos\delta' \tag{2.35}$$

$$U = I\,\cos 2\chi\,\sin 2\psi = 2\,\langle E_{x_0}\,E_{y_0} \rangle\,\cos\delta = 2\,\langle E_{l_0}\,E_{r_0} \rangle\,\sin\delta' \tag{2.36}$$

$$V = I\,\sin 2\chi \quad = 2\,\langle E_{x_0}\,E_{y_0} \rangle\,\sin\delta = \langle E_{l_0}^2 \rangle - \langle E_{r_0}^2 \rangle. \tag{2.37}$$

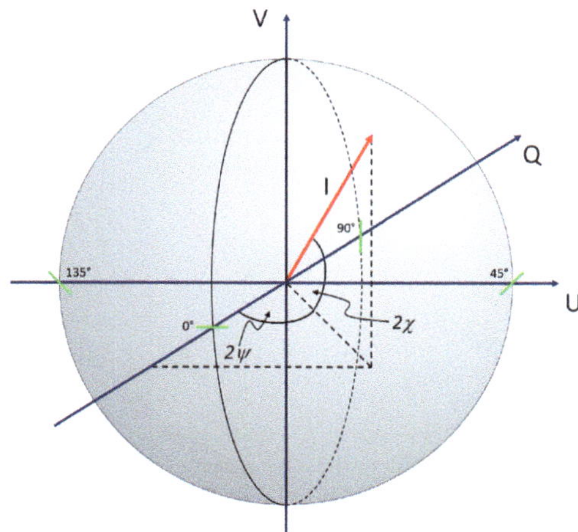

Fig. 2.26 Poincaré sphere

In the above equations, the brackets $<>$ represent time averages. It is evident that one can retrieve the relevant polarisation properties by either evaluating the differences $E_x - E_y$ or $E_l - E_r$, or by (cross-)correlating the electro-magnetic waves. The most reliable quantity is obtained from correlation, since in such an operation

$$< P > = \lim_{T \to \infty} \frac{1}{T} \int_{-\infty}^{+\infty} E_1(t)\, E_2(t)\, dt,$$

which is the quantity delivered by the outputs of the so-called IF polarimeters that are attached to the receivers of radio telescopes (see Sect. 3.2).[7]

In Figs. 2.27 and 2.28 examples of measured linear polarisation of synchrotron radiation are shown. In the first example we see the synchrotron radiation from the supernova remnant (SNR) 3 C10, which is a relic of the supernova observed by Tycho Brahe in 1572. The radio contours indicate increasing intensity when their arrows go anti-clockwise, and vice versa. Inspection of the left panel of the image shows that the SNR has a shell-type radio structure. The superimposed 'vectors' indicate the orientation of the magnetic field. Their length is proportional to the polarised intensity. What we see here is the synchrotron radiation from the relativistic electrons in this SNR. The right panel just shows the 'B-vectors', this time with their length proportional to the degree of polarisation, as defined by Eq. (2.29) (for more details on supernova remnants, see Sect. 4.3). Fractions

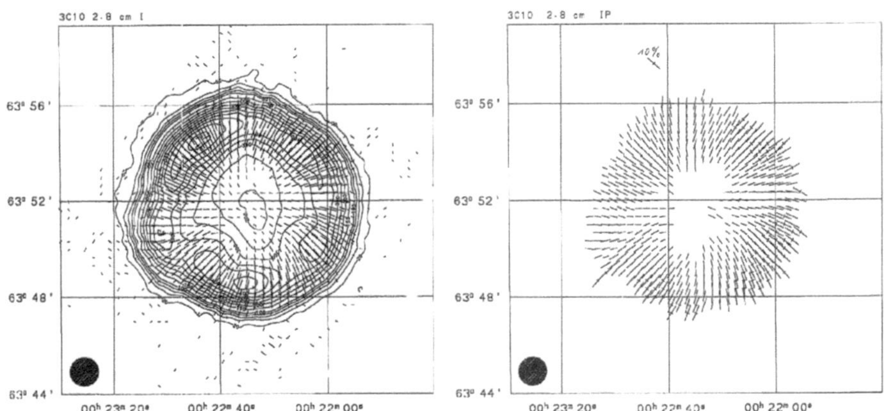

Fig. 2.27 Radio emission of the SNR 3C 10, with the magnetic-field orientation indicated by the bars (observed by one of the authors (UK) at 10.7 GHz using the Effelsberg 100-m telescope; see text for explanatory details)

---

[7]See textbooks on radio astronomy.

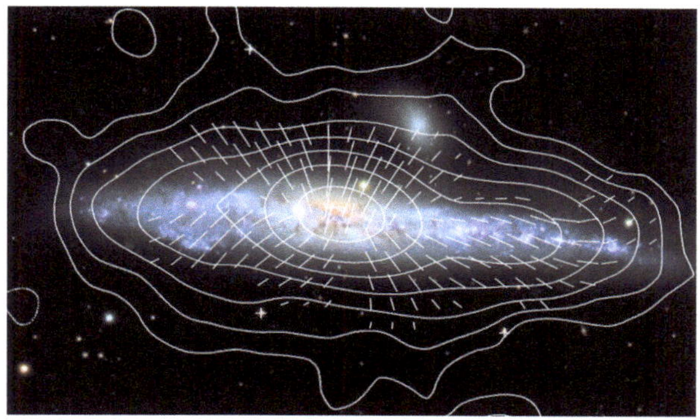

Fig. 2.28 Radio emission of the spiral galaxy NGC 4631 at 8.35 GHz (contours), superimposed onto an optical image. The *white bars* indicate the orientation of the magnetic field (Krause, 2009)

of polarisation exceeding 10 % are indicated by orthogonal dashes crossing the 'vectors'. The second example shows the radio emission of the spiral galaxy NGC 4631, which we happen to view 'edge on'. The radio emission (contours) is a mixture of free-free and synchrotron. Again, the magnetic field as traced by the linear polarisation is represented by 'B-vectors' (for more details on galaxies, see Chap. 5).

2.3.4 Losses and Particle Lifetimes

Synchrotron radiation causes energy losses of the radiating particles (electrons or positrons), which gives rise to a cutoff in their energy spectra above a certain energy, which in turn causes a steepening of the radiation spectra above some frequency. As we shall see, this signature in the spectra can be used as a tool to estimate the 'ages' of the particles, meaning the time elapsed since their energy replenishment ceased. In what follows we derive a relation between the energy and particle half-lifetime as a function of the strength of the magnetic field.

The energy losses due to synchrotron radiation correspond to the radiated power of the particles, which is

$$P = \frac{2}{3} \cdot \frac{e^2}{c^3} \cdot \dot{v}^2 \cdot \left(\frac{E}{m_0\, c^2} \right)^4$$

as we have seen. The acceleration can be written as

$$\dot{v} = \frac{v^2}{r_L} = \omega_L v,$$

and hence with

$$\omega_L = \frac{e\,B}{m_0\,c}\,\frac{1}{\gamma}$$

$$\dot{v} = \frac{v\,e\,B}{m_0\,c} \cdot \frac{m_0\,c^2}{E}.$$

Since $v \approx c$, we obtain

$$P = \frac{2}{3} \cdot \frac{e^4}{m_0^4\,c^7} \cdot B^2\,E^2.$$

Now, since

$$P \stackrel{!}{=} -\frac{dE}{dt}$$

it follows that

$$\frac{dE}{dt} = -2.37 \cdot 10^{-3} \cdot \left(\frac{B}{G}\right)^2 \cdot \left(\frac{E}{\mathrm{erg}}\right)^2 \mathrm{erg\,s}^{-1}$$

$$= -1.48 \cdot 10^{-3} \cdot \left(\frac{B}{\mu G}\right)^2 \cdot \left(\frac{E}{\mathrm{eV}}\right)^2 \mathrm{eV\,s}^{-1}.$$

We rewrite this in the form

$$\frac{dE}{E^2} = -a \cdot B^2 \cdot dt,$$

where $a = 2.37 \cdot 10^{-3}$ if E has units of erg, which upon integration yields

$$\frac{1}{E} - \frac{1}{E_0} = a \cdot B^2 \cdot (t - t_0).$$

E_0 corresponds to the initial energy at time t_0 that the particles have before the losses set in. The half-lifetime $t_{1/2}$ is defined as the time after which the particle has lost half its energy, i.e.

$$E(t_{1/2}) = \frac{E_0}{2}.$$

Taking $t_0 = 0$, we have

$$E(t) = \frac{1}{\frac{1}{E_0} + a \cdot B^2 \cdot t} = \frac{E_0}{1 + a \cdot B^2 \cdot E_0 \cdot t}$$

so that

$$\frac{E_0}{2} = \frac{E_0}{1 + a \cdot B^2 \cdot E_0 \cdot t_{\frac{1}{2}}},$$

or

$$t_{\frac{1}{2}} = \frac{1}{a} \cdot B^{-2} \cdot E_0^{-1}.$$

Inserting a and expressing the magnetic-field strength B in μG and the energy E in GeV, we arrive at

$$t_{1/2} = 8.35 \cdot 10^9 \cdot \left(\frac{B}{\mu G}\right)^{-2} \cdot \left(\frac{E_0}{\text{GeV}}\right)^{-1} \quad \text{year.} \tag{2.38}$$

Assuming a magnetic-field strength $B = 10\,\mu$G, as is typical for the ISM in the Milky-Way and in external galaxies, we find half-lifetimes of order $10^7 \dots 10^8$ year for particles with GeV energies (Table 2.3).

The evolution of a radio source is hence such that once the energy supply has been switched off, the energy spectrum will have a cutoff beyond some critical energy E_c that gradually migrates towards lower energies – initially rather rapidly at the highest energies, and progressively slower as the cutoff moves towards lower ones. This is sketched in the left panel of Fig. 2.29. As a result, the synchrotron radiation spectrum exhibits a corresponding exponential decline beyond the cutoff frequency ν_c, which we obtain from

$$\nu_c = \frac{3}{4\,\pi} \cdot \frac{e}{m_0^3 c^5} \cdot B\,E_c^2.$$

This cutoff frequency ν_c beyond which the source is rendered undetectable tells us something about the age of the source. Strictly speaking, this is rather the duration of the 'remnant phase', or the time elapsed since the energy source had been switched off, but it is commonly used as the *source age* synonymously. Using convenient units, it can be written as

$$\tau_{source} = 1.06 \cdot 10^9 \cdot \left(\frac{B}{\mu G}\right)^{-\frac{3}{2}} \cdot \left(\frac{\nu}{\text{GHz}}\right)^{-\frac{1}{2}} \quad \text{year.} \tag{2.39}$$

Table 2.3 Particle half-lifetimes ($B = 10\,\mu$G)

E (GeV)	$t_{1/2}$ (year)
1	$8 \cdot 10^7$
10	$8 \cdot 10^6$
100	$8 \cdot 10^5$

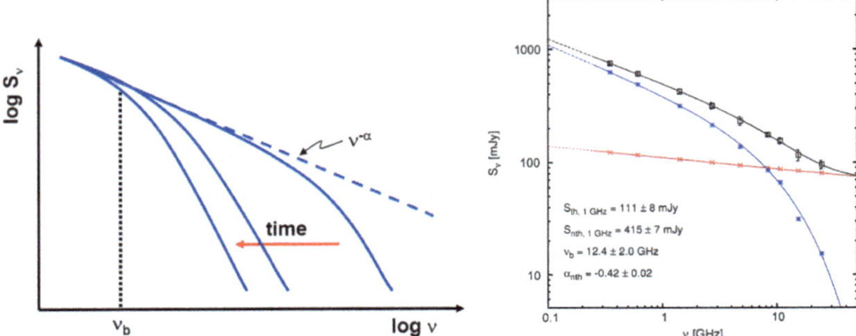

Fig. 2.29 *Left*: Migration of the cutoff frequency resulting from particle 'ageing'. *Right*: Radio continuum spectrum of the starburst dwarf galaxy NGC 1569, exhibiting a break in the synchrotron spectrum above $\sim$12 GHz

The right panel of Fig. 2.29 shows an example: the radio continuum spectrum of the starburst dwarf galaxy (see Sect. 5.3) NGC 1569 is shown here, with least-squares fits to the synchrotron (blue) and total (black) radio continuum spectrum. The measured flux densities (black squares) have been used, and the thermal free-free component (red) subtracted at each frequency, such as to yield the pure nonthermal synchrotron component (blue stars). The only fixed parameter in this fitting procedure is the spectral-index of the thermal radiation, viz. $\alpha_{th} = -0.1$, which is well defined by theory (see Sect. 2.2). Obviously in this case, one can 'date' the termination of the starburst producing the relativistic particles to about 5 Myr using Eq. 2.3.4, with a magnetic-field strength of about 20 μG inferred from the measured synchrotron intensity (see Sect. 3.2).

2.4 Inverse-Compton Radiation

Another process causing energy losses of relativistic particles is the so-called inverse-Compton process. In Compton scattering, a photon transfers energy and momentum to a free electron. If the free electrons have a sufficiently high kinetic energy, $m\,c^2 \gg h\,v$, the reverse may happen, i.e. net energy will be transferred from the electron to the photon. Relativistic electrons moving in a photon field, e.g. the far-infrared radiation from galaxies, or the ubiquitous micro-wave background (CMB) radiation, may 'boost' such photons into the X-ray (or even γ-ray) regime. The derivation of the radiated power is very similar to that for synchrotron radiation, hence we only present the result here:

$$\int \frac{dP_{IC}}{d\Omega}\, d\Omega = P_{IC} = \frac{4}{3} \cdot \sigma_T\, c\, \beta^2\, \gamma^2\, u_{rad}, \tag{2.40}$$

where σ_T is the Thomson cross section and

$$u_{rad} = \frac{4\,\sigma}{c}\,T^4$$

is the energy density of the radiation field, with σ the Stefan-Boltzman constant. We can express the radiated synchrotron power in an analogous form,

$$P_{syn} = \frac{2\,e^2}{3\,c^3}\cdot \dot{v}_{\perp}^2\,\gamma^4,$$

where

$$\dot{v}_{\perp} = \frac{e\,v_{\perp}\,B}{\gamma\,m_0\,c}.$$

Hence

$$P_{syn} = \frac{2\,e^4}{3\,c^3\,m_0^2}\cdot \beta_{\perp}^2\,\gamma^2\,B^2,$$

where we need to average over $\beta_{\perp}^2$. With $\beta_{\perp} = \beta\,\sin\chi$, we obtain

$$\langle\beta_{\perp}^2\rangle = \frac{\beta^2}{4\,\pi}\int_0^{2\pi}\!\!\int_0^{\pi}\sin^2\chi\,d\Omega = \frac{\beta^2}{4\,\pi}\int_0^{2\pi}\!\!\int_0^{\pi}\sin^3\chi\,d\chi\,d\psi = \frac{2}{3}\cdot\beta^2.$$

Thus, we obtain

$$P_{syn} = \frac{4\,e^4}{9\,m_0^2\,c^3}\cdot\beta^2\,\gamma^2\,B^2 = \frac{32\,\pi\,e^4}{9\,m_0^2\,c^3}\cdot\beta^2\,\gamma^2\cdot\frac{B^2}{8\,\pi} \qquad (2.41)$$

$$= \frac{4}{3}\cdot\sigma_T\,c\,\beta^2\,\gamma^2\,u_{mag}, \qquad (2.42)$$

where we have used the Thomson cross section and the classical electron radius,

$$\sigma_T = \frac{8\,\pi}{3}r_e^2 = 6.65\cdot 10^{-25}\,\mathrm{cm}^2$$

$$r_e = \frac{e^2}{m_0\,c^2} = 2.82\cdot 10^{-13}\,\mathrm{cm}.$$

We realise that the radiated power due to both, the synchrotron and the inverse-Compton process have identical dependencies, viz. on the square of the particle energy and on the square of the energy density of a photon or magnetic field.

As an example, let us assume that a synchrotron source (e.g. radio galaxy) is immersed in the CMB, which has $T_0 = 2.728\,\mathrm{K}$. We now calculate the ratio of both radiation powers by dividing (2.40) by (2.42). which is nothing but the ratio of the energy densities of these fields, i.e. that of the magnetic energy density and that of the energy density of the background-photon field (which is a perfect black body):

$$\frac{P_{syn}}{P_{IC}} \overset{!}{=} \frac{u_{mag}}{u_{rad}} = \frac{B^2}{8\,\pi} \cdot \frac{c}{4\,\sigma} \cdot T^{-4},$$

where the CMB temperature depends on redshift:

$$T = T_0 \cdot (1 + z).$$

We can now ascribe this temperature an equivalent magnetic field B_{eq} of the CMB, i.e.:

$$\frac{B_{eq}^2}{8\,\pi} = \frac{4\,\sigma}{c} \cdot T_0^4 \, (1 + z)^4,$$

so that for $T_0 = 2.728\,\mathrm{K}$ we finally obtain an equivalent magnetic field strength of the CMB of

$$B_{eq} = \sqrt{\frac{32\,\pi\,\sigma}{c}} \cdot T_0^2 \, (1 + z)^2$$
$$= 3.25 \, (1 + z)^2 \, \mu\mathrm{G} \,.$$

A relativistic particle moving in such an environment 'does not care' whether it loses energy via synchrotron or inverse-Compton, because of the identical dependence on their energy and on the energy density of the magnetic and the photon fields. Accounting for both loss mechanisms reduces, of course, the particle lifetime resulting in

$$\boxed{\; t_{1/2} = 1.59 \cdot 10^9 \cdot \frac{B^{\frac{1}{2}}}{B^2 + B_{eq}^2} \cdot \left[\left(\frac{\nu}{\mathrm{GHz}}\right) \cdot (1 + z)\right]^{-\frac{1}{2}} \; \text{year}, \qquad (2.43)\;}$$

where B and B_{eq} are in $\mu\mathrm{G}$. Note that the numerical value in this equation differs from that in (2.38) by a factor of 1.5, since we have averaged here over the pitch angles of the radiating electrons. One of the first examples in which inverse-Compton radiation caused by the CMB was detected is the radio galaxy 3 C294. This source exhibits extended inverse-Compton emission (Fig. 2.30). Its redshift is $z = 1.779$, yielding an equivalent magnetic field of $B_{eq} \approx 27\,\mu\mathrm{G}$! The interpretation of the observed structures is quite intriguing: while along the current jet axes (see Sect. 7.4 for the taxonomy and features of AGN) fresh relativistic particles are transported into the radio lobes, producing synchrotron radiation there (white

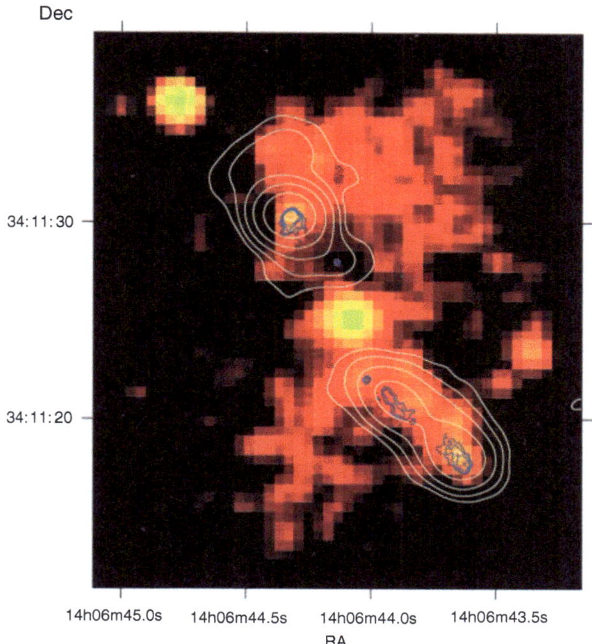

Fig. 2.30 Synchrotron (contours) and inverse-Compton (colour) radiation from the radio galaxy 3C 294 (Erlund et al. 2006)

contours), inverse-Compton radiation is seen in the X-rays (colour), produced by particles that have cooled significantly over the past, but still able to boost the CMB photons to X-ray energies. The implication is that the radio jets are precessing, which is a not too rare phenomenon, indicating the presence of binary supermassive black holes. The aged particles in the precessed jets and lobes should be detectable at low radio frequencies, e.g. with LOFAR.

We will come back to the inverse-Compton process when dealing with radio galaxies (Sect. 7.4) and clusters of galaxies (Sect. 8.6). In general, one can calculate the equivalent inverse-Compton field strength in the following way. If L_{bol} is the bolometric luminosity of a region of size d, the intensity is given by the flux density S and the solid angle Ω, so that for a distance D this reads:

$$I = \frac{S}{\Omega} = \frac{L}{4\pi\, D^2\, \Omega}\,.\tag{2.44}$$

Plugging in the solid angle for a (circular) region of size d, the distance cancels out and

$$I = \frac{L}{\pi^2\, d^2}\,.\tag{2.45}$$

The radiation energy density

$$u_{rad} = \frac{4\pi}{c} I \, , \tag{2.46}$$

is then equated to the magnetic energy density

$$u_{mag} = \frac{B^2}{8\pi} \, , \tag{2.47}$$

which leads to

$$B_{eq} = \sqrt{\frac{32}{c}} \cdot \frac{L^{1/2}}{d} \, . \tag{2.48}$$

The bolometric luminosity can be obtained from far-infrared measurements, making use of the fact that it is well approximated by $L \approx \nu \cdot L_\nu$. So, for example, if one measures the 24 μm flux densities for a number of HII regions in galaxies (Kennicutt et al., 2007), one obtains $L = 10^{41}$ erg s^{-1} for HII regions in M 51 and aperture sizes of 13″ (hence, $d \approx 500$ pc). Plugging this into Eq. (2.48), one obtains $B_{eq} \sim 7 \, \mu$G. The strength of the magnetic field in M 51 is $B \approx 15 \, \mu$G (Sect. 5). So, at least for the spiral-arm regime inverse-Compton losses are significant and need to be considered when calculating particle lifetimes.

2.5 Spinning Dust Grains

In the course of studies of the Galactic foreground emission as a potential contaminant of the CMB anisotropy studies, 'anomalous' radio emission was discovered at frequencies above $\sim$15 GHz that could not be attributed to synchrotron, free-free, or thermal (vibrational) emission from dust. Clarification came witch a seminal paper by Draine and Lazarian (1998) on radio emission from spinning dust grains. This electric dipole radiation is produced by rotationally excited, small interstellar dust grains producing a bump in the radio continuum spectrum in the range 14 ... 50 GHz. In Fig. 2.31 the radio continuum spectrum of the starburst galaxy M 82 is shown, with the thermal free-free and synchrotron components as discussed in Sects. 2.2, 2.3, and 2.3.4. At frequencies above about 100 GHz the thermal radiation from dust (not discussed in this book) dominates the overall luminosity of galaxies. It is obvious from this spectrum that one can measure the thermal free-free radiation alone only within a small frequency range between about 40 and 70 GHz (a frequency range that is not easily accessible from the ground, owing to the atmospheric attenuation). It is this very range where one would expect the radiation emitted by spinning dust grains to be seen. The *expected*(!)[8] bump produced by

---

[8]Note that the detection of emission from spinning dust grains has remained controversial up to now. The bump shown here has not been measured yet, as this requires utter precision of the measurements!

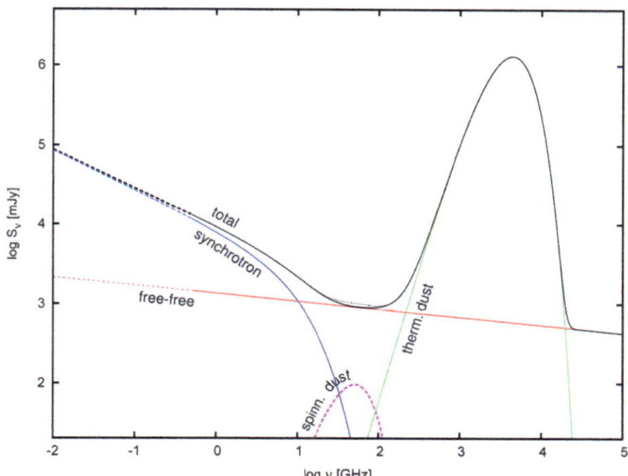

Fig. 2.31 Superposition of various continuum radiation components in the radio spectrum of M 82

this kind of radiation is indicated in Fig. 2.31 by the dashed magenta line, with the contribution to the total (solid black line) emission represented by the dotted black line, which is barely visible in the frequency range of lowest overall emissivity of this galaxy.

The details of calculating the frequency spectrum of spinning dust grains are complex, the present book not providing enough space for this. We will hence just have a look at the general route to get there. A spinning grain with electric dipole moment $\vec{\mu}$ radiates a power

$$P = \frac{2}{3} \cdot \frac{\omega^4 \, \mu^2 \, \sin^2 \Theta}{c^3},$$

which if averaged over Θ average of

$$\langle P \rangle_\Theta = \frac{4}{9} \cdot \frac{\mu^2 \, \omega^4}{c^3}.$$

Here, θ is the angle between the angular velocity $\vec{\omega}$ and the moment $\vec{\mu}$. In order to calculate the emissivity, one has to consider various damping and excitation mechanisms controlling the spins of the grains – a tedious task, and in spite of the seminal paper quoted above probably not yet delivering a matured theoretical concept. In fact, all of the detections claimed to date have not been all that convincing, also because of the difficulty to separate the radio emission from spinning dust from the other emission components (synchrotron and free-free). The net spins of the dust particles depend on the balance between damping and excitation of their rotation.

For *damping*, we need to consider:

- Collisional drag: species like H, H_2, and He may stick on the surface when hitting a grain, and may be desorbed with a thermal velocity distribution.

- Plasma drag: a grain's electric dipole moment $\vec{\mu}$ interacts with passing ions, hence it couples the grain rotation to the plasma. This process is the dominant damping mechanism in molecular clouds.
- Infrared emission: a thermally excited photon of energy $h\,\nu$ released from a grain with $\omega \ll \nu$ removes, on average, $\hbar\omega/2\pi\nu$ of angular momentum from the grain. This is the dominant process in the warm neutral medium (WNM).
- Electric dipole damping: this accounts for the radiated power P.

For *excitation*, we need to consider:

- Recoil from thermal collisions and evaporation: neutrals and ions deposit their angular momentum when they impact the grain, and give it an additional kick when they subsequently evaporate. This bombardment by neutrals and ions is the dominant process for small grains in molecular clouds, in the cold neutral medium (CNM), in the WNM, and in the WIM.
- Excitation by the plasma: this process implies the excitation of the grain rotation by the fluctuating electric field of a passing plasma, and is at work in molecular clouds and for large grains.
- Infrared emission: each IR photon carries away an angular momentum $\hbar$, corresponding to a change in angular momentum of the grain.
- Photo-electric emission: photo-electrons are emitted randomly from the grains. Since $m_e \ll m_H$, this mechanism is negligible.

In order to derive the emissivity, one has to account for the excitation and damping rates. These depend on the temperature T and on the size distribution dn/da of the dust particles. The emissivity per H atom finally reads:

$$\frac{j_\nu}{n_H} = \left(\frac{8}{3\pi}\right)^{\frac{1}{2}} \cdot \frac{1}{n_H c^3} \int da\, \frac{dn}{da} \cdot \frac{\mu^2\,\omega^6}{\langle \omega^2 \rangle^{\frac{3}{2}}} \cdot e^{-\frac{3\omega^2}{2\langle \omega^2 \rangle}}. \tag{2.49}$$

This emissivity gives rise to an emission bump noticeable in the frequency range $15 \le \nu \le 50\,\mathrm{GHz}$. Because of the superposition of the various continuum emissions in this frequency domain, its separation from other emission components is anything but easy (see Fig. 2.31).

References

Draine, B.T., Lazarian, A., 1998, Astroph. J. **508**, 157
Erlund, M.C., Fabian, A.C., Blundell, K.M., et al., 2006, MNRAS **371**, 29
Fomalont, E.B., Ebneter, K.A., van Breugel, W.J.M., Ekers, R.D., 1989, Astroph. J. **346**, 17
Haslam, C.G.T., Salter, C.J., Stoffel, H., Wilson, W.E., 1982, Astron. Astroph. **47**, 1
Haynes, R.F., Klein, U., Wayte, S.R., et al., 1991, Astron. Astroph. **252**, 475
Kennicutt, R.C., Calzetii, D., Walter, F., et al., 2007, Astroph. J. **671**, 333
Kim, S. Staveley-Smith, L., Dopita, M.A., et al. 1998, Astroph. J. **503**, 674
Krause, M., 2009, Rev. Mex. Astron. Astrof. **36**, 25
Longair, M.S., 2011, High-Energy Astrophysics, Cambr. Univ. Press, Chapt. 8

Schwinger, J., 1949, Phys. Rev. **75**, 1912
Swordy, S., 1995, unpublished compilation, Univ. Chicago
Wallis, G., 1959, in *Paris Symposium on Radio Astronomy*, ed. R.N. Bracewell, p. 595
Wills, K.A., Pedlar, A., Muxlow, T.W.B., Wilkinson, P.N., 1997, MNRAS **291**, 517
Wilson, T.L., Pauls, T., 1984, Astron. Astroph. **138**, 225

Chapter 3
Diagnostics

In this chapter, we shall treat the main diagnostic tools that we have at hand to investigate interstellar and intergalactic magnetic fields. There are essentially four different tracers of magnetic fields in a diffuse astrophysical medium:

- Optical polarisation, caused by dust grains whose magnetic moments align along the magnetic field.
- Synchrotron radiation, bearing information about the total strength of the magnetic field. Its linearly polarised component measures the strength and orientation of the ordered or uniform component of the magnetic field perpendicular to the line of sight.
- Faraday rotation, yielding information about the line-of-sight component of the magnetic field *and its direction.*
- The Zeeman effect, yielding information about strong magnetic fields in dense gas clouds and about their direction.

In what follows, the physics of these diagnostic tools is discussed, the observables pointed out, and examples of measurements are given. The emphasis will be on synchrotron radiation and Faraday rotation, because their measurement constitutes the least challenge of all methods described here and because the interpretation of the measurements is relatively straightforward. In addition to the techniques listed above, a fifth tool will be briefly described at the end of this chapter, namely the polarised radiation from dust and molecules. The required measurements are difficult, and – in contrast to the other methods – their interpretation may be sophisticated.

3.1 Optical Polarisation

Optical (stellar) light is polarised by selective extinction, due to elongated dust particles that are aligned with the interstellar magnetic field, owing to their magnetic moments (Fig. 3.1). The extinction is strongest along the dust particles' major

© Springer International Publishing Switzerland 2015
U. Klein, A. Fletcher, *Galactic and Intergalactic Magnetic Fields*,
Springer Praxis Books, DOI 10.1007/978-3-319-08942-3_3

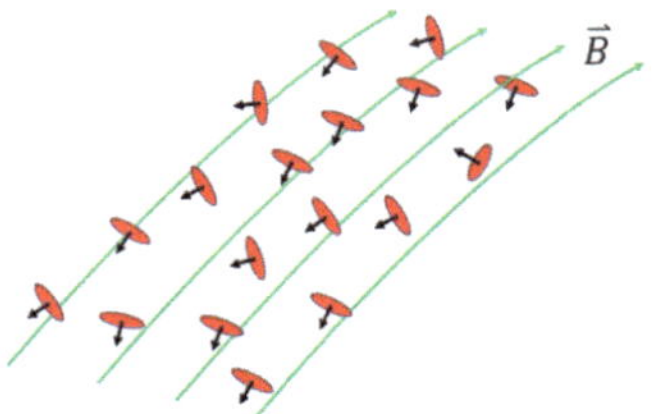

Fig. 3.1 Elongated grains lining up in a magnetic field

axes, assuming that they have a prolate structure. Hence, the resulting net (linear) polarisation has $\vec{E} \parallel \vec{B}$, which is different from synchrotron radiation where we saw that $\vec{E} \perp \vec{B}$. The extinction cross-section is

$$C_{ext} = C_{abs} + C_{sca} = \pi\, a^2\, Q_{ext},$$

where a is the typical radius of dust grains, and Q_{ext} is the so-called extinction efficiency,

$$Q_{ext} = Q_{abs} + Q_{sca}.$$

It is related to the albedo

$$\gamma = \frac{Q_{sca}}{Q_{ext}} = \frac{Q_{sca}}{Q_{abs} + Q_{sca}},$$

where $\gamma = 1$ implies full reflection, and $\gamma = 0$ implies no scattering at all. More realistically, the dust particles have a size distribution $f(a)$ so that

$$C_{ext}(\lambda) = \int_0^\infty Q_{ext}(a, \lambda) \cdot \pi\, a^2\, f(a)\, da.$$

For silicates and graphites, $f(a) \propto a^{-3.5}$. The resulting absorption coefficient is

$$\kappa_\lambda = \pi\, a^2 \cdot Q_{ext} \cdot n_d,$$

where n_d is the number density of dust particles. We define the extinction cross sections $C_\parallel$ and $C_\perp$ for the electric field parallel and perpendicular to dust particles' longitudinal axes, respectively. Then, the corresponding intensities can be written as

$$I_\parallel = \frac{1}{2}\, I_0 \cdot e^{-\tau_\parallel}, \qquad I_\perp = \frac{1}{2}\, I_0 \cdot e^{-\tau_\perp}$$

with

$$I = I_\parallel + I_\perp.$$

The magnitude of the polarisation then is

$$\Delta m_p = -2.5 \cdot \log \frac{I_\parallel}{I_\perp} = 1.086 \cdot (\tau_\parallel - \tau_\perp),$$

where

$$\tau_\parallel = C_\parallel \cdot n_d \cdot r, \qquad \tau_\perp = C_\perp \cdot n_d \cdot r.$$

In general, the alignment of the dust particles is not perfect. Hence, one introduces the degree of alignment $f \in [0, 1]$ via

$$\Delta m_p = 1.086\, f \cdot (C_\parallel - C_\perp)\, n_d\, r.$$

The geometry of rotating dust particles exposed to an interstellar magnetic field is illustrated in Fig. 3.2. Dust particles are likely to be non-spherical. Otherwise, anisotropic extinction would not be possible. Let us consider them as prolate (rods or ellipsoids). Because of their continuous collisions with gas particles (mainly hydrogen atoms), the dust particles are permanently forced to rotate. In thermal equilibrium, the rotational energy around longitudinal and transverse axis is equal, i.e.

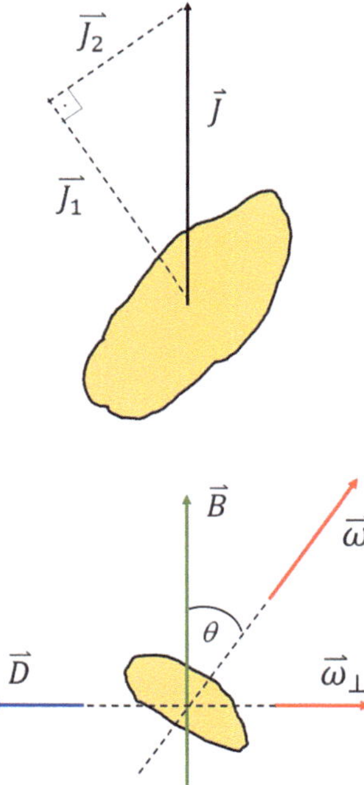

Fig. 3.2 Geometry of rotating grain and magnetic field

$$E_{rot,i} = \frac{1}{2} \cdot I_i\, \omega_i^2 = \frac{1}{2} \cdot k\, T_g.$$

Here, T_g is the gas temperature, I_i is the moment of inertia of the dust particle, and the index i refers to the principle axes of the elongated particle. Since $I_i\, \omega_i^2 = K$ is approximately constant, we have $\omega_i \sim 1/\sqrt{I_i}$, hence its angular momentum can be expressed as

$$J_i = I_i \cdot \omega_i = K\,\sqrt{I_i}.$$

Since usually $I_1 \gg I_2$, the total angular momentum $\vec{J} = \vec{J}_1 + \vec{J}_2$ will have its vector close to $\vec{J}_1$, i.e. along the transverse axis of the dust particle. Hence the rotation will primarily be perpendicular to the symmetry axis. A rough estimate of ω_i can be made in the following way: consider a spherical particle with radius a and density ρ_d, hence mass

$$m_d = \frac{4}{3}\, \pi\, a^3\, \rho_d\,.$$

Then

$$I = \frac{2}{5} \cdot m_d\, a^2 = \frac{8}{15}\, \pi\, a^5\, \rho_d\,,$$

where

$$\frac{1}{2}\, I\omega^2 = \frac{1}{2}\, k\, T_g\,.$$

Plugging in $a = 5 \cdot 10^{-5}\,\mathrm{cm}$, $\rho_d = 1\,\mathrm{g\,cm^{-3}}$, $T_g = 100\,\mathrm{K}$, this yields $\omega = 2 \cdot 10^4\,\mathrm{rad\,s^{-1}}$! The colliding particles are hydrogen atoms in the first place. Many of them stick to the dust grain surface, combining to H_2 with another hydrogen atom. The particles thereby get rid of the excess energy in this exothermic process (dust is a catalyst in H_2 formation!), thereby transferring extra energy to the dust grain. Since the surface of the grain is not uniform, this will cause a net increase of its angular momentum on average. An estimate leads to an increase by a factor of about 10^4 so that $\omega \approx 10^8 \ldots 10^9\,\mathrm{rad\,s^{-1}}$!

The alignment of the particles was worked out by Davis and Greenstein (1951) and explained in terms of paramagnetic relaxation. If the rotation axis of the particle does not coincide with the direction of the magnetic field, each of its volume elements is exposed to a periodically varying magnetic field and, correspondingly, to a varying magnetisation $\vec{M}$ per unit volume (Fig. 3.2). The direction of $\vec{M}$ permanently deviates from the instantaneous direction of $\vec{B}$, lagging behind the latter by some amount. This behaviour is described in terms of a complex magnetic susceptibility $\chi = \chi' + i\,\chi''$, where the imaginary part determines the portion of rotational energy that is continuously converted into heat by re-magnetisation.

In other words: it is associated with an additional magnetisation $\vec{M}''$, which is perpendicular to $\vec{B}$ and $\vec{J}$. One finds

$$\vec{M}'' = \frac{\chi''}{\omega} \cdot \vec{\omega} \times \vec{B}.$$

This gives rise to a torque

$$\vec{D} = V \cdot \vec{M}'' \times \vec{B}$$

with modulus

$$D = V \cdot |\vec{M}''| \cdot |\vec{B}| \sin \Theta = V \, \chi'' \, B^2 \, \sin \Theta,$$

where V is the grain's volume and the angle $\Theta = \angle(\vec{\omega}, \vec{B})$. The torque $\vec{D}$ is directed perpendicular to $\vec{B}$ and has an opposite direction to $\omega_\perp$, the component of $\vec{\omega}$ perpendicular to $\vec{B}$. The angular momentum $I \cdot \omega_\perp$ around the axis perpendicular to $\vec{B}$ is gradually diminished, finally vanishing after a retardation time given by

$$D \cdot t_{ret} = I \cdot \omega_\perp.$$

What remains is the rotation about an axis parallel to $\vec{B}$, while the magnetisation does not change anymore. For many paramagnetic substances the relation

$$\chi'' \approx 2.5 \cdot 10^{-12} \cdot \frac{\omega_\perp}{T_d}$$

holds, where T_d is the dust temperature. An estimate of the retardation time $t_{ret} = I \, \omega_\perp / D$ yields $\sim 4 \cdot 10^7$ year, assuming $a = 5 \cdot 10^{-5}$ cm, $\rho_d = 1 \, \mathrm{g \, cm^{-3}}$, $T_d = 15 \, \mathrm{K}$, $B = 3 \, \mu\mathrm{G}$, and $\sin \Theta = 0.5$. Now one has to compare this with the time t_c required to significantly change the rotation of a dust grain by collisions, which can be estimated by assuming that this is the time required to have collisions with particles making up for the total mass of the dust grain. For this we need N collisions

$$N_{coll} = \frac{m_d}{m_H} = t_{coll} \, v_{coll}.$$

Here, the collision frequency is

$$v_{coll} = \pi \, a^2 \, v_g \, n_H,$$

where

$$v_g = \sqrt{\frac{8 \, k \, T_g}{\pi \, m_H}}. \tag{3.1}$$

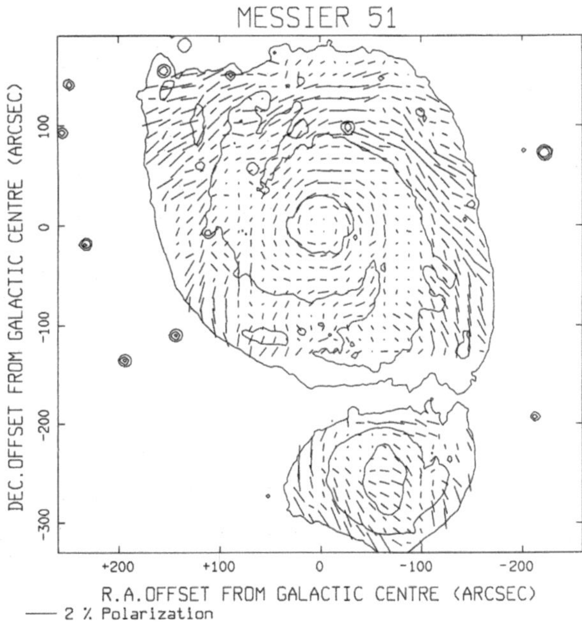

Fig. 3.3 Optical polarisation of the spiral galaxy M 51 (Scarrott et al., 1987)

Using the previous parameters ($a = 5 \cdot 10^{-5}\,\text{cm}$, $T_g = 100\,\text{K}$, $n_H = 1\,\text{cm}^{-3}$), one arrives at $t_{coll} \lesssim 10^6\,\text{year}$, which is much shorter than t_{ret}. This means that but few dust grains reach perfect alignment. A 'precise' derivation of the degree of alignment delivers

$$f \approx 0.3 \cdot \left(\frac{t_{coll}}{t_{ret}} \right) \sim B^2 \cdot \left(\frac{\chi''}{\omega} \right) \cdot \frac{1}{a\, n_H\, \sqrt{T_g}}\,.$$

Hence, in order to obtain sizeable degrees of alignment, strong magnetic fields are required. Yet, the polarisation of starlight has been readily observed in the general ISM, in the presence of magnetic fields of several μG strength, a nice example shown in Fig. 3.3. The magnetic field in the spiral galaxy M 51 appears to closely follow the prominent spiral structure of this galaxy, and is thus in accord with what is seen in the radio domain (compare with Fig. 5.1). So, since starlight polarisation is readily observed, it looks as if the theory has not been completely understood to date.

3.2 Synchrotron Radiation

3.2.1 Polarisation

Synchrotron radiation is (partly) polarised, and hence provides a powerful tool to study cosmic magnetic fields whenever and wherever charged relativistic particles (electrons, positrons) are present. As we have seen in Sect. 2.3.3, the polarisation state of an electromagnetic wave is elegantly described by the so-called Stokes parameters I, U, Q and V, which define the Poincaré sphere. The total power carried by the radiation (or, depending on the measured quantity, the intensity, flux density, ...) is represented by Stokes I, such that

$$I^2 = Q^2 + U^2 + V^2. \tag{3.2}$$

As we have also seen, the Stokes parameters can either be expressed in terms of parameters defining the polarisation ellipse, or as combinations of the Cartesian or circular components (Eqs. 2.34–2.37). By means of so-called IF polarimeters,[1] radio astronomers can relatively easily and reliably measure all Stokes parameters, from which one can then deduce the total polarisation state of the observed electromagnetic wave. As we shall see in Sect. 3.3.2, the actually measured degrees of polarisation are always much lower than the theoretically predicted ones, owing to all kinds of depolarisation mechanisms. One therefore makes a distinction between the measured total intensity, given by Stokes I,

$$I_{tot} = I, \tag{3.3}$$

and the measured polarised intensity as retrieved from

$$I_{pol} = \sqrt{U^2 + Q^2 + V^2}. \tag{3.4}$$

In many astrophysical applications we are dealing with linear polarisation, hence $V = 0$, so that in this case

$$I_{pol} = \sqrt{U^2 + Q^2}. \tag{3.5}$$

The degree of polarisation is in general given by

$$p_{tot} = \frac{\sqrt{U^2 + Q^2 + V^2}}{I}, \tag{3.6}$$

and the degree of linear polarisation by

$$p_{lin} = \frac{\sqrt{U^2 + Q^2}}{I}. \tag{3.7}$$

---

[1] The IF (intermediate frequency) polarimeter is a so-called quadrature network that allows to measure the Stokes parameters by correlating the orthogonal components E_x, E_y or E_l, E_r such as to yield the desired quantities of Eqs. (2.34)–(2.37).

In this case, we also need to measure the polarisation angle ψ, which we obviously retrieve from (Eqs. 2.35 and 2.36), i.e.

$$\psi = \frac{1}{2} \arctan \frac{U}{Q}. \tag{3.8}$$

For sources with circular polarisation, the degree is obviously

$$p_{circ} = \frac{V}{I}.$$

Normal synchrotron sources (SNR, diffuse emission from the MW and from galaxies, from radio galaxies, or from clusters of galaxies) emit linearly polarised radio emission, while circularly polarised radiation is seen in sources emitting gyro-cyclotron or coherent emission (the Sun, Jupiter, pulsars, cosmic masers). In what follows, we shall briefly discuss the physical quantities that can be deduced from measurements of synchrotron radiation. The intensity of synchrotron radiation is (Sect. 2.3.2)

$$I_\nu \sim B_\perp^{1+\alpha} \cdot \nu^{-\alpha},$$

where

$$\alpha = \frac{g-1}{2},$$

α being the slope of the power-law radiation spectrum, and g that of the energy spectrum of relativistic particles,

$$N(E)\, dE = A\, E^{-g}\, dE.$$

Since radio waves traverse the universe essentially unhindered by matter, we can derive the shape of the particle spectrum of synchrotron-emitting sources out to large distances. The spectral break seen at high frequencies helps to estimate the particle age, i.e. the time elapsed since these particles have been injected or accelerated (Sect. 2.3.4).

3.2.2   Equipartition Field Strength

The brightness of the synchrotron radiation furthermore allows to estimate the total strength of the magnetic field, or more precisely, that of $B_\perp$. In order to do so, we have to assume energy equipartion between the relativistic particles and the magnetic field in which they are gyrating. The radio luminosity is

$$P_\nu = c' \, A \, V \, B^{1+\alpha} \, \nu^{-\alpha} = P_{\nu_0} \cdot \left(\frac{\nu}{\nu_0}\right)^{-\alpha}, \tag{3.9}$$

where V is the volume of the source and c' a constant. The total energy of the source responsible for the radio emission is the sum of the energies of the particles and the magnetic field:

$$E_{tot} = E_{part} + E_{mag} = V \cdot \int_0^\infty E \, N(E) \, dE + \frac{B^2}{8\,\pi} \cdot V.$$

At this point, we have to take into account the energy of the relativistic protons in this balance. Even though they do not radiate significantly, they make up for the bulk of the particle energy. We account for this by writing

$$E_p = \beta \, E_e$$

and hence for the total particle energy

$$E_{part} = (1 + \beta) \cdot E_e = \eta \, E_e$$

we therefore write

$$E_{tot} = \eta \, V \cdot \int_0^\infty N(E) \, E \, dE + V \cdot \frac{B^2}{8\,\pi}.$$

Recalling the relation between the critical frequency of the radio emission and the strength of the magnetic field,

$$\nu = \frac{3}{4\,\pi} \cdot \frac{e\,B}{m_0\,c} \cdot \left(\frac{E}{m_0\,c^2}\right)^2 = C \, E^2 \, B$$

we can write the boundaries of the above integral, i.e. the minimum and maximum energies of particles in terms of the corresponding frequencies of the emission:

$$E_{min,max} = \left(\frac{\nu_{min,max}}{C\,B}\right)^{\frac{1}{2}},$$

for which the radio spectrum is known. It is generally assumed that $\nu_{min} \approx 10\,\mathrm{MHz}$, and $\nu_{max} = 100\,\mathrm{GHz}$, which are the extremes of the (classical) radio regime. Even though both of these limits are somewhat arbitrary, one can argue: the lowest-energy particles[2] contribute little to the total energy budget. In normal radio sources with declining power-law spectra, particles producing radio emission above the upper frequency bound are unlikely. We can then calculate the total energy contained in the relativistic particles via

$$E_{part} = \eta\, V \cdot \int_{E_{min}}^{E_{max}} E\, A\, E^{-g}\, dE = \frac{\eta\, V\, A}{g-2} \cdot \left[\left(\frac{\nu_{min}}{C\,B}\right)^{\frac{2-g}{2}} - \left(\frac{\nu_{max}}{C\,B}\right)^{\frac{2-g}{2}} \right].$$

Solving (3.9) for A and substituting this into the above equation, we have

$$E_{part} = \frac{\eta\, V}{g-2} \cdot \left(\frac{P_\nu}{c'\, V\, B^{1+\alpha}\, \nu^{-\alpha}} \right) \cdot (C\,B)^{\frac{g-2}{2}} \cdot \left[\nu_{min}^{(2-g)/2} - \nu_{max}^{(2-g)/2} \right].$$

Preserving only the essential observables, we arrive at

$$E_{part} = G \cdot \eta \cdot P_\nu \cdot B^{-\frac{3}{2}}, \tag{3.10}$$

where G is a constant that depends weekly on α, ν_{min}, and ν_{max}. Therefore, the total energy is

$$E_{tot} = G\, \eta\, P_\nu\, B^{-\frac{3}{2}} + V \cdot \frac{B^2}{8\,\pi}.$$

In Fig. 3.4, a plot of the magnetic-field, particle, and total energy is shown, with a minimum total energy for a magnetic field $B_{min} = B(E_{min})$, which is near to the magnetic field strength producing precise equipartition, $B_{eq} = B(E_{eq})$. Minimizing this with respect to B, we obtain

$$B_{min} = \left(6\,\pi \frac{G\,\eta\,P_\nu}{V} \right)^{\frac{2}{7}}. \tag{3.11}$$

Substituting B_{min} into (3.10), we obtain

$$E_{mag} = V \cdot \frac{B^2}{8\,\pi} = \frac{3}{4} \cdot E_{part}.$$

We realise that the condition for the minimum-energy requirement to produce the observed radio emission from a given radio source corresponds very closely to a condition of equipartition of energy between particles and magnetic fields. The minimum total energy is

---

[2]Even though the low-energy CR spectrum is strongly modulated by the solar wind as mentioned before.

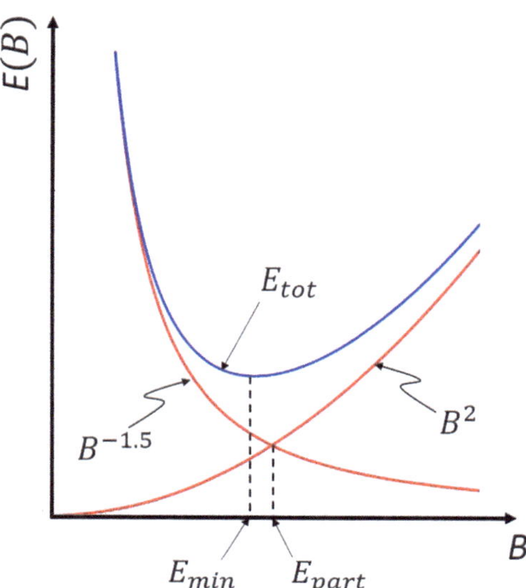

Fig. 3.4 Minimum-energy and equipartition magnetic field strengths

$$E_{tot,min} = \frac{7}{24\,\pi}\, V^{\frac{3}{7}} \cdot (6\,\pi\, G\, \eta\, P_\nu)^{\frac{4}{7}}$$

A useful approximation for this if we plug $\alpha = 0.75$ (a typical value) into the quantity G and assuming a spherical volume of radius R is

$$E_{min} = 6\cdot 10^{41}\cdot\left(\frac{\nu}{100\ \mathrm{MHz}}\right)^{\frac{2}{7}}\cdot\left(\frac{P_\nu}{\mathrm{W\,Hz}^{-1}}\right)^{\frac{4}{7}}\cdot\left(\frac{R}{\mathrm{kpc}}\right)^{\frac{9}{7}}\cdot\eta^{\frac{4}{7}}\ \mathrm{erg}. \tag{3.12}$$

The ratio of energy carried by protons to that carried by electrons is of order $\eta = 10^2$. Move recent estimates, based on the theory of 'diffusive shock acceleration' (Sect. 4.4) yield $\eta = 20\ldots 40$. Assuming energy equipartition (which is not very different from minimum energy, which a system usually maintains), we can utilise the synchrotron intensity to infer the total magnetic-field strength. Even though this can only give rough estimates, it is a powerful method in the absence of any other diagnostic tools, such as Faraday rotation, or Zeeman measurements. We can then estimate the total magnetic-field strength locally by using the brightness temperature given by

$$S_\nu = \frac{2\,\nu^2\,k}{c^2}\cdot T_b\,\Omega_S,$$

where S_ν is the flux density and Ω_S the solid angle of the observed source, S_ν being related to the monochromatic radio luminosity (the radio power) via

$$S_\nu = \frac{P_\nu}{4\,\pi\,D^2},$$

where D is the distance to the source. For $h\nu \ll k T_b$, the Rayleigh-Jeans approximation has been utilised as can be generally assumed in the radio domain ($\nu < 50\,\text{GHz}$). Noting that the volume in (3.11) is $V \approx L^3$ (L = size of the source) and

$$\Omega_S = \left(\frac{L}{D}\right)^2 ,$$

one can derive the minimum-energy magnetic-field strength of a galaxy disk with radius L and effective 'radio thickness' H as

$$B_{min} = 1.7 \cdot \eta^{\frac{2}{7}} \cdot \left(\frac{T_b}{\text{K}}\right)^{\frac{2}{7}} \cdot \left(\frac{\nu}{\text{GHz}}\right)^{\frac{4}{7}} \cdot \left(\frac{H}{\text{kpc}}\right)^{-\frac{2}{7}} \mu\text{G} \tag{3.13}$$

so that, inserting typical values ($\eta = 40$, $T_b = 1\,\text{K}$, $\nu = 1\,\text{GHz}$, $H = 1\,\text{kpc}$) we obtain a value of $B_{min} \approx 5\,\mu\text{G}$. This is in line with independent determinations based upon measurements of Faraday rotation.

3.3 Faraday Rotation

3.3.1 Rotation Measure

In this section, we discuss the rotation of the electric-field vector that a linearly polarised wave experiences when propagating through a magnetised plasma, an effect called *Faraday rotation*. To this end, let us consider the propagation of electro-magnetic waves through a magnetised plasma. In order to treat this mathematically, we have to add the forces acting on the free electrons. If some force is exerted on a free charge in a magnetic field, the charged particles are forced into helical paths, owing to the Lorentz force. Hence we have

- Centrifugal force, $-m \cdot \omega^2 \cdot \vec{r}$
- The Lorentz force, $\pm \frac{e}{c} \cdot \omega \cdot B_\parallel \cdot \vec{r}$
- Electric force of propagating wave, $e \cdot \vec{E}(\vec{r}, t)$

A linearly polarised wave is a superposition of two circularly polarised waves with opposite rotation (LHC, RHC). $B_\parallel$ is the component of the magnetic field along (parallel to) the direction of propagation of the electro-magnetic wave. The geometry is illustrated in Fig. 3.5. The equation of motion then reads:

$$-m\,\omega^2 \vec{r} = \pm\frac{e}{c} \cdot \omega\, B_\parallel\, \vec{r} + e\, \vec{E}(\vec{r}, t) .$$

In the above, the '+' sign stands for a right-hand circularly polarised wave (RHC), and the '−' sign stands for a left-hand circularly polarised wave (LHC). Since the

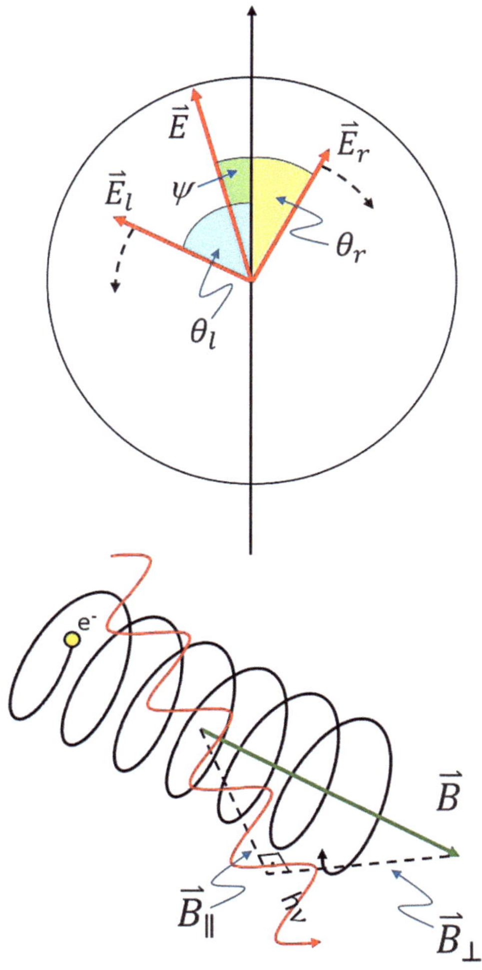

Fig. 3.5 Geometry of circularly and linearly polarised waves

RHC polarised wave has the same helicity as the electron, and the LHC polarised wave has the opposite helicity to it, this yields a difference in the interaction forces. This implies that if a circularly polarised wave has the same sense as the gyrating electron, the interaction is larger. The different interactions will have different effects on the phases of the RHC and LHC components of the wave. Solving the above equation for $\vec{r}$, we obtain

$$\vec{r} = -\frac{e}{m_0} \cdot \left(\frac{1}{\omega^2 \pm \omega\,\omega_c} \right) \cdot \vec{E} \; ,$$

where

$$\omega_c = \frac{e\,B_\parallel}{m_0\,c}$$

is the cyclotron frequency. The force acting on the electron implies an *electric displacement* $\vec{D}$, described by the *electric polarisation*

$$\vec{P} = \frac{1}{4\pi}\cdot(\vec{D}-\vec{E}) = \frac{\epsilon-1}{4\pi}\cdot\vec{E} = n_e\,e\,\vec{r}$$

$$\Longrightarrow -n_e\cdot\frac{e^2}{m_0}\cdot\frac{1}{\omega^2\pm\omega\cdot\omega_c} = \frac{1}{4\pi}\cdot(\epsilon-1)\,,$$

or

$$\epsilon = 1 - \frac{\omega_p^2}{\omega\,(\omega\pm\omega_c)} \qquad \omega_p = \sqrt{\frac{4\,\pi\,n_e\,e^2}{m_0}}\,.$$

The quantity ω_p is the plasma frequency, which we encountered in the previous section. Recalling that the refractive index is $n = \sqrt{\epsilon}$ (ϵ is the dielectric constant) we arrive at a general expression for the refractive index, which accounts for the presence of a magnetic field (contained in the cyclotron frequency ω_c):

$$n = \sqrt{1 - \frac{\omega_p^2}{\omega\,(\omega\pm\omega_c)}}\,.$$

The difference to the refractive index for $|\vec{B}| = 0$ is that it implies different phase velocities of the RHC component E_r and the LHC component E_l of the wave, resulting in a phase shift of E_r with respect to E_l after traversing the magnetised plasma. This phase shift between E_r and E_l implies a rotation of the linearly polarised electric field of the electromagnetic wave. Referring to the illustration on the previous page, the reader should be clear about the following: in order for the electric field vector $\vec{E}$ to rotate by a certain amount, its individual RHC and LHC components have to rotate by twice this angle. Considering the rotation of the electric-field vector along some path dr through the magnetised plasma and denoting the corresponding rotation of the RHC component by $d\theta_r$ and that of the LHC component by $d\theta_l$, we calculate the net rotation $d(\Delta\psi)$ along the path dr as we have

$$d\theta_l = k_l\cdot dr$$

$$d\theta_r = k_r\cdot dr$$

$$d(\Delta\psi) = \frac{k_l - k_r}{2}\cdot dr\,.$$

The phase velocity is

$$v_p = \frac{\omega}{k} = \frac{c}{n} \implies k = \frac{\omega}{c} \cdot n$$

so that

$$d(\Delta\psi) = \frac{\Delta k}{2} \cdot dr = \frac{\omega}{2\,c} \cdot \Delta n \, dr.$$

With $n = \sqrt{\epsilon}$, we can calculate the corresponding difference of the refractive index,

$$n_l^2 - n_r^2 = \epsilon_l - \epsilon_r = (n_l + n_r) \cdot (n_l - n_r) = 2n\,\Delta n \qquad (n_l \approx n_r)$$

$$\implies \Delta n = \frac{\epsilon_l - \epsilon_r}{2 \cdot n}$$

Inserting here ϵ_l, ϵ_r yields

$$\Delta n = \frac{\omega_p^2\,\omega_c}{n\,\omega} \cdot \frac{1}{\omega^2 - \omega_c^2} \ .$$

The refractive index n was given by

$$n = \left[1 - \frac{\omega_p^2}{\omega\,(\omega \pm \omega_c)} \right]^{\frac{1}{2}} ,$$

which for $\omega \gg \omega_c$ and $\omega \gg \omega_p$ is approximated by

$$n \approx 1 - \frac{1}{2} \cdot \left(\frac{\omega_p}{\omega} \right)^2 \ .$$

For the difference of the refractive index of the RHC and the LHC components this yields

$$\Delta n = \frac{\omega_p^2\,\omega_c}{\omega^3} \ .$$

This change of phase Δn is translated into a change in polarisation angle $\Delta\psi$ of the linearly polarised wave penetrating the magnetised plasma via

$$d(\Delta\psi) = \frac{\omega_p^2\,\omega_c}{2\,c\,\omega^2}\,dr \ .$$

Finally, the total Faraday rotation of the electric vector is obtained by integrating this equation over the total path length

$$\Delta\psi = \frac{2\,\pi\,e^3}{m_0^2\,c^2\,\omega^2} \cdot \int_0^{r_0} n_e\,B_\parallel\,dr\,.$$

Conventionally, this is written in the form

$$\Delta\psi = RM \cdot \lambda^2\,, \qquad (3.14)$$

where *RM* is the so-called *rotation measure*. This important quantity is measured in units of rad m^{-2}, the wavelength λ being measured in m. In astrophysical applications, the rotation measure is written such that it contains convenient units, i.e.

$$RM = 0.81 \cdot \int_0^{r_0} \left(\frac{n_e}{cm^{-3}}\right) \cdot \left(\frac{B_\parallel}{\mu G}\right) \cdot \left(\frac{dr}{pc}\right)\ rad\ m^{-2}\,. \qquad (3.15)$$

Its dependence on λ^2 tells us that Faraday rotation decreases rather quickly as we go to higher radio frequencies. At the same time, this makes it a very powerful tool as a probe of the interstellar (and intergalactic) magnetic field. If *RM* can be determined by measuring the Faraday rotation angle $\Delta\psi$ over a sufficient frequency range, then we have information about the interstellar magnetic-field strength if n_e is known. The determination of RM requires measurements of ψ at least three frequencies, owing to an ambiguity to measure ψ. With just two measurements, the rotation could be $\Delta\psi + n \cdot \pi$, where n is an integer.

Below, we list some examples of rotation measures in different environments relevant for astronomy. An example of the distribution of the rotation measure across a spiral galaxy is shown in Fig. 3.6 (see Chap. 5 for more details on magnetic fields in galaxies). The absolute values of the rotation measure in external galaxies are typically $|RM| \lesssim 100$ rad m^{-2}, while in some radio galaxies they can be very high, $|RM| \lesssim 5000$ rad m^{-2}. The reason for the large values in such sources (like e.g. in Hydra A, see Fig. 8.5) is that these objects reside in the centres of cooling-flow galaxy clusters, so that we are dealing with a large line-of-sight through a magnetised and fully ionised region.

$$\text{Ionosphere:} \quad \left.\begin{array}{l} n_e \approx 10^6\ cm^{-3} \\ r_0 \approx 100\ km \\ B_\parallel \approx 0.3\ G \end{array}\right\} \quad RM \approx 0.8\ rad\ m^{-2}$$

$$\text{ISM:} \quad \left.\begin{array}{l} n_e \approx 0.03\ cm^{-3} \\ r_0 \approx 500\ pc \\ B_\parallel \approx 5\ \mu G \end{array}\right\} \quad RM \approx 60\ rad\ m^{-2}$$

$$\text{Cooling flows:} \quad \left.\begin{array}{l} n_e \approx 0.005\ cm^{-3} \\ r_0 \lesssim 200\ kpc \\ B_\parallel \approx 3\ \mu G \end{array}\right\} \quad RM \lesssim 3000\ rad\ m^{-2}$$

Fig. 3.6 Observed RM between 3.5 and 6.2 cm wavelength across the spiral galaxy NGC 6946 (Beck et al. 2007)

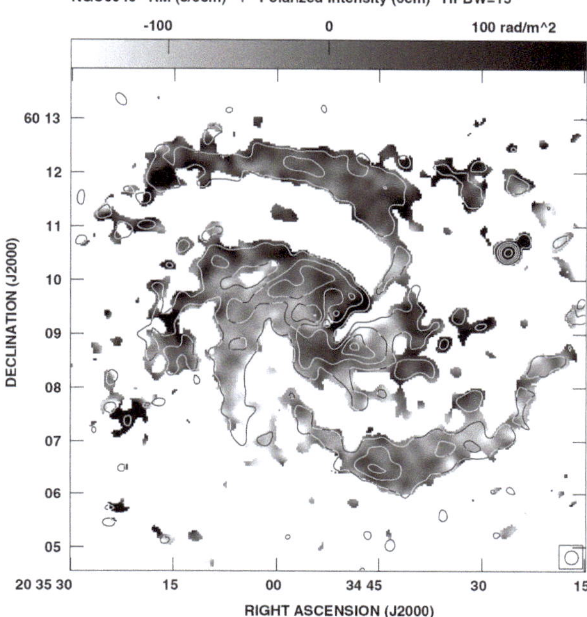

Fig. 3.7 RM ambiguity: measured values (*blue*) can be shifted by multiples of π (*red*) such as to produce a different RM

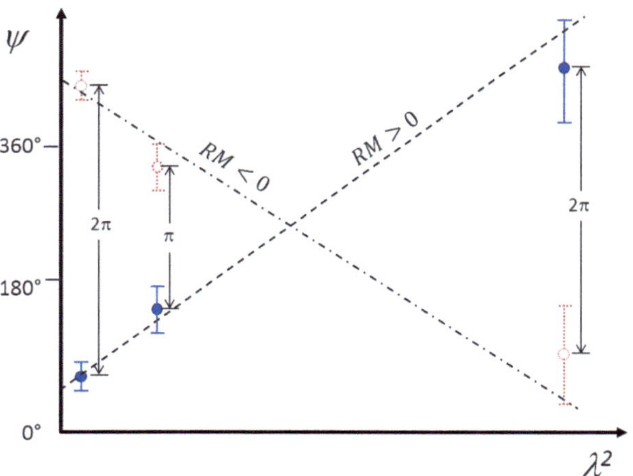

It is difficult to derive the rotation measure unambiguously, since one cannot distinguish between ψ and $\psi + n\,\pi$.[3] The sketch in Fig. 3.7 shows a possible least-squares fit to measured polarisation angles $\psi(\lambda^2)$ that allow for both, a positive and

---

[3]The reason for this is that the response of a dipole to a linearly polarised electromagnetic wave is identical when the dipole is rotated by $180°$, and any detector of linear polarisation is subject to this principle!

negative *RM* (or even larger absolute values) if too few wavelengths are involved in the measurements! The newly developed rotation-measure synthesis technique used at lower radio frequencies (Sect. 3.3.8) is a very promising tool to resolve this problem. Otherwise, the recipe is to observe the target at as many wavelengths as possible.

3.3.2   *Depolarisation*

Astrophysical objects with a particle energy spectrum of the form $N(E)\,dE \sim E^{-g}\,dE$ produce radio emission with a flux density $S_\nu \sim \nu^{-\alpha}$, where $\alpha = (g-1)/2$. In this case, we expect (Sect. 2.3.3) a 'theoretical' degree of linear polarisation

$$p = \frac{g + 1}{g + \frac{7}{3}} = \frac{\alpha + 1}{\alpha + \frac{5}{3}},$$

which of course may vary with location, i.e. $p = p(\vec{r})$. This degree of polarisation depends only weakly on frequency. Its maximum is, however, never reached, owing to a number of effects that reduce this degree of polarisation, commonly referred to as depolarisation. We shall see, however, that both, Faraday rotation as well as depolarisation deliver information about the magnetised plasmas. If the instrumental effects are known, the ratio of the observed to the theoretical degree of polarisation bears astrophysical information that can then be utilised.

An instrumental effect that reduces the degree of depolarisation is the so-called bandwidth depolarisation. The reason for this is that the polarisation angle rotates over the bandwidth (Fig. 3.8) in the presence of Faraday rotation. If the bandpass has limiting wavelengths λ_1 and λ_2, then the electric-field vector will be rotated across that band according to

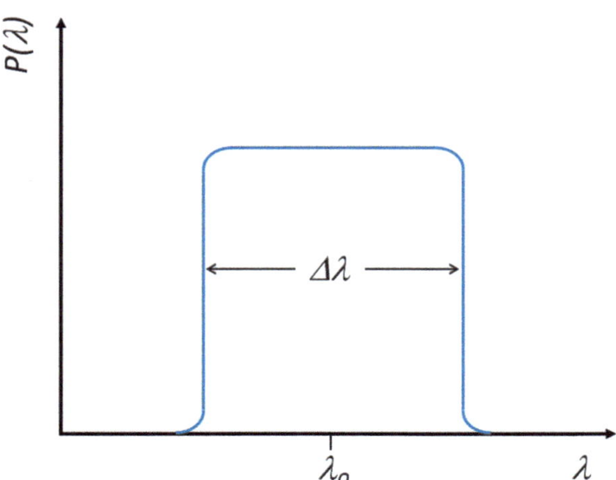

Fig. 3.8 Bandpass

$$\Delta\psi = RM \cdot (\lambda_1^2 - \lambda_2^2)$$

$$\approx (\lambda_1 + \lambda_2) \cdot (\lambda_1 - \lambda_2) \cdot RM$$

$$\approx 2\,\lambda_0\,\Delta\lambda \cdot RM,$$

where we have made use of

$$\lambda = \frac{c}{\nu} \Rightarrow \frac{d\lambda}{d\nu} = -\frac{c}{\nu^2} \Rightarrow \Delta\lambda = -c\,\frac{\Delta\nu}{\nu_0^2}$$

and hence

$$\Delta\psi = -2\,\lambda_0^2 \cdot RM \cdot \frac{\Delta\nu}{\nu_0}.$$

We shall see below how this rotation across the bandpass leads to a reduction of the degree of linear polarisation, given a particular rotation measure.

The degree of polarisation is generally obtained by integrating the degree of polarisation $P(\vec{r})$ over some path r, weighted by the emissivity $\epsilon(\lambda,\vec{r})$, and neglecting any frequency dependence of $P(\vec{r})$:

$$P_{obs}(\lambda) = \frac{\iint\limits_{source} p(\vec{r}) \cdot \epsilon(\lambda,\vec{r}) \cdot e^{i\,2\,[\psi_0(\vec{r})+\lambda^2\Phi(\vec{r})]}\,dr\,d\Omega}{\iint\limits_{source} \epsilon(\lambda,\vec{r})\,dr\,d\Omega}. \tag{3.16}$$

Here,

$$P(\vec{r}) = p(\vec{r}) \cdot e^{i\,2\,\psi_0(\vec{r})}$$

is the intrinsic degree of polarisation of the source. Note that the degree of polarisation defined in this way is a complex quantity![4] If the polarised emission comes from a region within which the Faraday rotation varies, it is convenient to define the rotation or Faraday depth $\Phi(\vec{r})$ (Fig. 3.9) for the rotation measure appropriate to each volume element at position $\vec{r}$:

$$\Phi(\vec{r}) = K \cdot \int\limits_0^{\vec{r}} n_e\,\vec{B} \cdot d\vec{r}. \tag{3.17}$$

Here K is the constant preceding the integral in Eq. (3.15), its numerical value depending on the units used for the thermal electron density, the magnetic-field strength, and the path length. Equation (3.16) can now be simplified by superposing

---

[4]Note that we use upper-case P for the complex polarisation and lower-case p for its modulus!

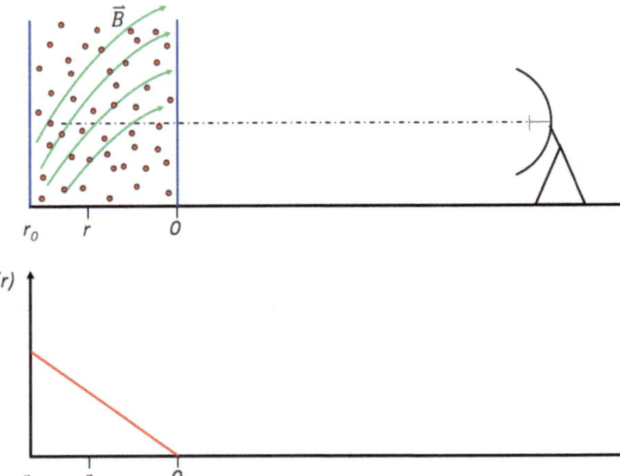

Fig. 3.9 Illustration of Faraday depth

all radiation with the same Faraday or rotation depth. This simultaneously accounts for lateral variations. If $E(\Phi)\,d\Phi$ is the fraction of total emission (assumed independent of λ) with a rotation depth between Φ and $\Phi + d\Phi$, and if the intrinsic polarisation is

$$P(\Phi) = p(\Phi) \cdot e^{i\,2\,\psi(\Phi)},$$

then the polarisation at wavelength λ is

$$P(\lambda^2) = \int_{-\infty}^{+\infty} E(\Phi)\,P(\Phi)\,e^{i\,2\,\Phi\lambda^2}\,d\Phi = \int_{-\infty}^{+\infty} F(\Phi)\,e^{i\,2\,\Phi\lambda^2}\,d\Phi, \tag{3.18}$$

where the source function $F(\Phi) = E(\Phi) \cdot P(\Phi)$, also called Faraday dispersion function, is the polarised intensity, expressed as a fraction of the total intensity of the source. Obviously, $P(\lambda^2)$ is the Fourier transform of $F(\Phi)$. Unfortunately, we cannot simply invert this Fourier integral, such as to obtain

$$F(\Phi) = \frac{1}{\pi} \cdot \int_{-\infty}^{+\infty} P(\lambda^2) \cdot e^{-i\,2\,\Phi\lambda^2}\,d(\lambda^2),$$

as $P(\lambda^2)$ is not defined for $\lambda^2 < 0$. We shall, however, see how we can retrieve $P(\lambda^2)$ by assuming simple model distributions for $F(\Phi)$ and perform the Fourier inversion (3.18). The resulting degree of polarisation may be compared with the observations, and parameters of the dispersion function $F(\Phi)$ adjusted until best agreement with the observations is achieved (s.b.). Recently, however, a technique

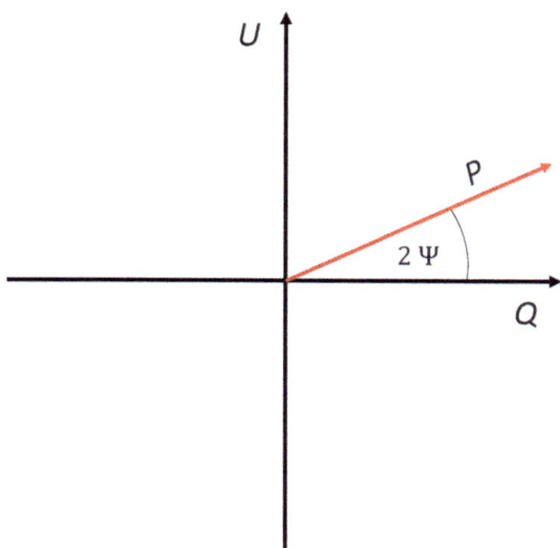

Fig. 3.10 The complex (Q,U)-plane

has been developed that allows to do the Fourier inversion inspite of the restrictions
inherent to it. This so-called 'RM synthesis technique' will be treated in Sect. 3.3.8.

Let us have a look at a few different situations, for which plausible forms of the
dispersion function can be assumed. For this method, which was first proposed in a
seminal paper by Burn (1966) it is convenient to work with the complex polarisation
defined by

$$P = Q + i\,U = p \cdot e^{i\,2\,\psi},$$

which describes the pseudo-vector P in the complex plane defined by the Stokes
parameters Q and U (see Fig. 3.10). Its magnitude is p, and its orientation is given
by *twice* the polarisation angle, in accord with Fig. 2.26 and Eqs. (2.34)–(2.37).
The Faraday dispersion function $F(\Phi)$ measures the polarised flux coming from
a Faraday depth Φ,

$$P(\lambda^2) = \int_{-\infty}^{+\infty} F(\Phi) \cdot e^{i\,2\,\Phi\,\lambda^2}\, d\Phi.$$

3.3.3 *Resolved Foreground Screen*

This is the trivial case, in which the Faraday-rotating screen has a uniform rotation
measure across the resolution element, i.e. the telescope's half-power beam width
(HPBW). In this case, the Faraday depth is constant across the beam, hence there is
no depolarisation, and just Faraday rotation:

$$\psi(\vec{r}) = \psi_0(\vec{r}) + \Phi(\vec{r}) \cdot \lambda^2,$$

where

$$\psi_0 = \psi(\lambda = 0)$$

is the intrinsic polarisation angle.

3.3.4   Unresolved Foreground Screen

In case of an unresolved foreground screen, the polarisation angle ψ does not rotate as a function of λ^2 anymore, since in this case there is a superposition of cells within the HPBW having different rotation depths (Fig. 3.11). Suppose we have random fluctuations of the magnetic field and/or electron density in a region extending a distance R from the observer. If the size scale d of the fluctuations is sufficiently small, i.e. $d \ll \theta_S \cdot R$, where θ_S is the angular diameter of the radio source, the Faraday dispersion function is well represented by a Gaussian with variance $K \langle n_e B_\| \rangle_c^2 d R$. Here $\langle n_e B_\| \rangle_c^2$ is the variance of the product of the electron density and the magnetic-field strength along the line-of-sight of a cell. The degree of polarisation then is

$$P(\lambda^2) = p_0 \cdot e^{-2 K^2 \langle n_e B_\| \rangle_c^2 d R \lambda^4}.$$

Figure 3.11 shows how the electric-field vectors are rotated by different amounts, depending on their line-of-sight to the observer. They correspondingly add up with a zig-zag pattern in the complex (Q,U)-plane.

The above equation rests upon the assumption that the dispersing cells fill the entire region. If we now assume a filling factor $\eta < 1$ for cells of dimension d with an average separation D, then the probability for having m clouds along line-of-sight is $\eta^m e^{-\eta} (m!)^{-1}$. In this case, the above equation takes the form

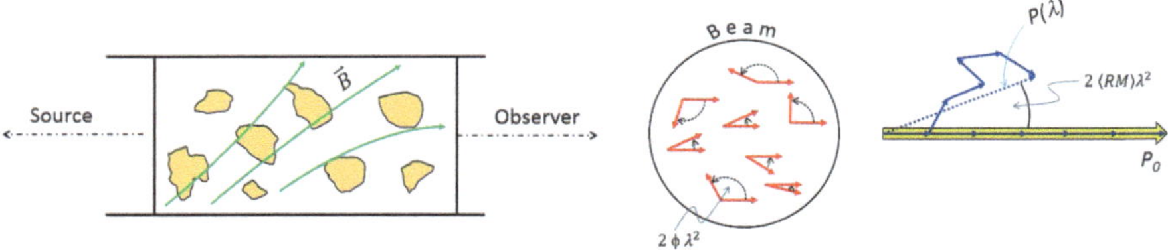

Fig. 3.11 Illustration of an unresolved foreground Faraday screen

$$P(\lambda^2) = p_0 \cdot exp\left[-\eta\left(1 - e^{-2\,K^2\,\langle n_e\,B_\parallel\rangle_c^2\,d^2\,\lambda^4}\right)\right],$$

where we have again assumed that $d \ll \theta_S \cdot R$. The quantity

$$\eta = \frac{d^2 \cdot R}{D^3}$$

is the average number of cells along the line-of-sight. What this equation tells us is that the fraction $e^{-\eta}$ not covered by cells is not affected while the rest is depolarised rather quickly.

3.3.5 Internal Faraday Depolarisation

In case of Faraday rotation happening within the source, things become more complicated, the resulting depolarisation being sketched in Fig. 3.12 for different geometries. The simplest geometry one can think of is a uniform slab, in which the linear path along the line-of-sight through the source is constant, assumed to be L here. In this case, Burn (1966) has shown that

$$P(\lambda^2) = p_0 \cdot \frac{1 - e^{-S}}{S},$$

where

$$S = (K\,n_e\,B_r)^2\,d\,L\,\lambda^4 - i\,2\,K\,n_e\,B_{u,\parallel}\,L\,\lambda^2.$$

Here, B_r stands for the isotropic random field, $B_{u,\parallel}$ denotes the line-of-sight component of the uniform field, n_e the electron density, d the size scale of the fluctuation, and L is the thickness of slab. In case of a uniform sphere, the above takes the form

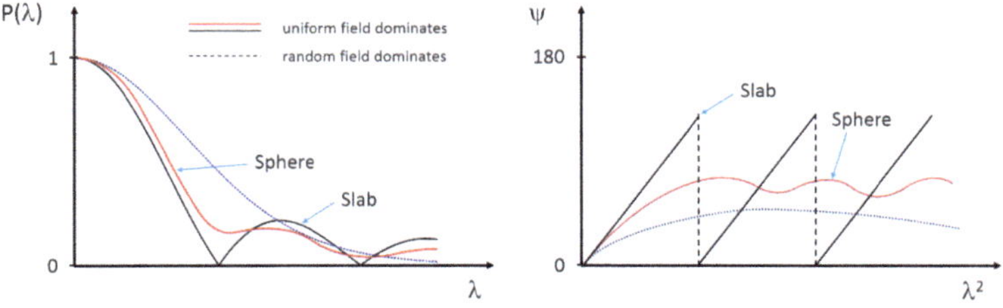

Fig. 3.12 Dependence of the degree of polarisation and polarisation angle on wavelength in case of internal depolarisation (after Gardner and Whiteoak, 1966)

$$P(\lambda^2) = p_0 \, \frac{3\left[(S+1)\,e^{-S} + \frac{1}{2}S^2 - 1\right]}{S^3},$$

where L is the diameter of the sphere. The properties of the above models depend on the ratio of the real to imaginary part of the quantity S in the wavelength range at which most of the depolarisation occurs, i.e. where $|S| \approx 1$. This is given by the number of cells along the line-of-sight $N = L/d$ through the slab and by the ratio of the energy densities in the uniform to total field strength $(B_r/2\,B_{u,\parallel})^2$. If the imaginary part of S dominates, i.e.

$$N \gg \left(\frac{B_r}{2 \cdot B_{u,\parallel}}\right)^2,$$

then we are dealing with a slab having a largely uniform magnetic field, which gives rise to a depolarisation law of the form

$$\boxed{P(\lambda) = p_0 \cdot \frac{\sin \Phi \, \lambda^2}{\Phi \, \lambda^2},} \qquad (3.19)$$

where

$$\Phi = K\, n_e\, B_{u,\parallel}\, L.$$

Obviously, Eq. (3.19) is a sinc function.[5] This is not a surprise, since the sinc function is the Fourier transform of a box function, which we have implicitly used to describe the simple uniform slab. One can now think of more complicated models, representing regions with different emission coefficients $\epsilon_1, \epsilon_2, \ldots$ and Faraday depths $\Phi_1, \Phi_2, \ldots$. Each of these will have depolarisations

$$\frac{\sin \lambda^2 \Phi_1}{\lambda^2 \Phi_1} \cdot e^{i\,\lambda^2\,\Phi_1}, \quad \frac{\sin \lambda^2 \Phi_2}{\lambda^2 \Phi_2} \cdot e^{i\,\lambda^2\,\Phi_2}, \ldots$$

If the intrinsic polarisation angle is constant (and set to zero for simplicity), the Stokes parameters defining the linear polarisation can be written by assigning the cosine terms of the exponential to Stokes Q and the sine terms to Stokes U:

$$Q = \sum_j \frac{\epsilon_j \sin \lambda^2 \Phi_j \cos \lambda^2 \Phi_j}{\lambda^2 \Phi_j} = \sum_j \frac{\epsilon_j \sin 2\,\lambda^2 \Phi_j}{2\,\lambda^2 \Phi_j}$$

$$U = \sum_j \frac{\epsilon_j \sin \lambda^2 \Phi_j \sin \lambda^2 \Phi_j}{\lambda^2 \Phi_j} = \sum_j \frac{\epsilon_j \sin^2 \lambda^2 \Phi_j}{2\lambda^2 \Phi_j}.$$

---

[5] $\mathrm{sinc}(x) := \frac{\sin \pi x}{\pi x}$.

The individual values of Q will be positive and negative, hence tend to cancel, while all values of U are positive. As a result, the zeros of the function $P(\lambda^2)$ disappear, and the resulting polarisation angle becomes $\langle \psi \rangle \approx 45°$, with damped oscillations about this value. Finally, we can state that if

$$N \gg \left(\frac{B_r}{2B_{u,\parallel}} \right)^2 ,$$

the imaginary part of S dominates and Faraday rotation is significant. On the other hand, if

$$N \ll \left(\frac{B_r}{2B_{u,\parallel}} \right)^2 ,$$

then the real part of S dominates, and the depolarisation goes along without any significant Faraday rotation as sketched in Fig. 3.12. In nature, there could be more sophisticated geometries involving different distributions of the thermal electrons (with number density n_e) and the synchrotron-emitting relativistic electrons (with number density N_e). This is sketched in Fig. 3.13. In Fig. 3.14 an example of depolarisation is shown. The synchrotron radiation from the galaxy cluster A 2256 (see Chap. 8 for magnetic fields in galaxy clusters) at 20 cm wavelength is shown. The central region is characterised by a diffuse 'halo', which is completely depolarised, while the peripheral 'relic' is strongly polarised, most likely because of its location away from, and in front of, the central cluster gas.

3.3.6 Beam Depolarisation

Assume that we have uniform field component B_u, with a random component B_r superimposed. We are then faced with the phenomenon of 'beam depolarisation' such that (Burn, 1966)

$$p_{obs} = p(g) \cdot \frac{B_u^2}{B_u^2 + B_r^2} \qquad (3.20)$$

is observed, where (see Sect. 2.3.3)

$$p(g) = \frac{g+1}{g+\frac{7}{3}} .$$

The simple reason for beam depolarisation is that different field orientations within the HPBW give rise to different polarisation angles which then partly cancel. Note that linearly polarised emission from two regions with magnetic-field

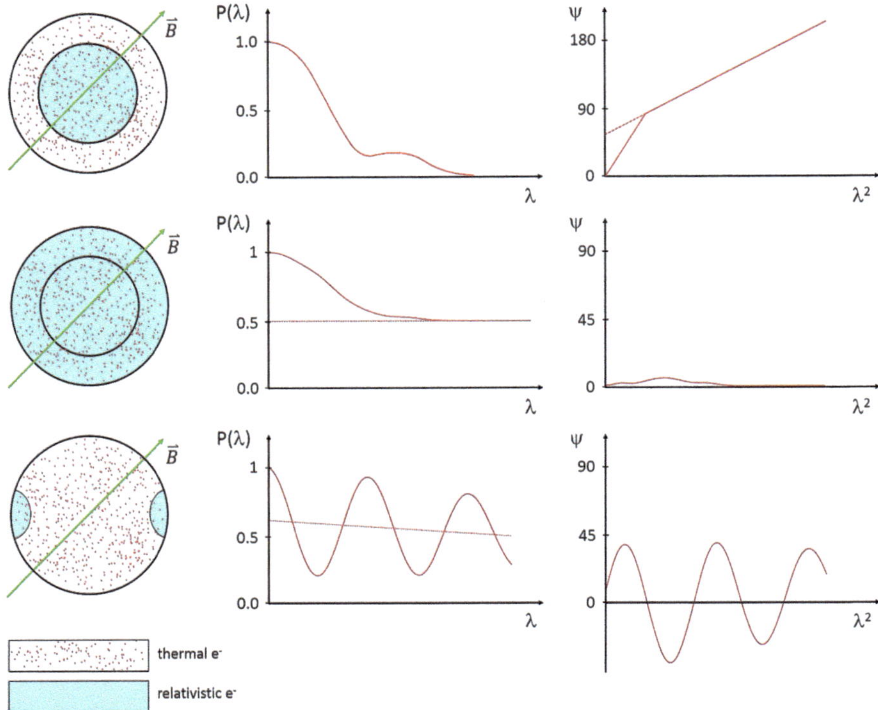

Fig. 3.13 Internal depolarisation for three idealised source models (after Gardner and Whiteoak, 1966)

orientations perpendicular to each other cancels out. The reason is that a difference of 90° of the polarisation angle in the sky corresponds to a rotation by 180° in the (Q, U)-plane.

3.3.7   Bandwidth Depolarisation

This effect can be calculated using a slab with a constant dispersion function $F(\Phi)$ between 0 and Φ_0:

$$\frac{p(\lambda^2)}{p(0)} = \frac{\displaystyle\int_0^{\Phi_0} e^{i\,2\,\lambda^2\,\Phi}\,d\Phi}{\displaystyle\int_0^{\Phi_0} d\Phi}$$

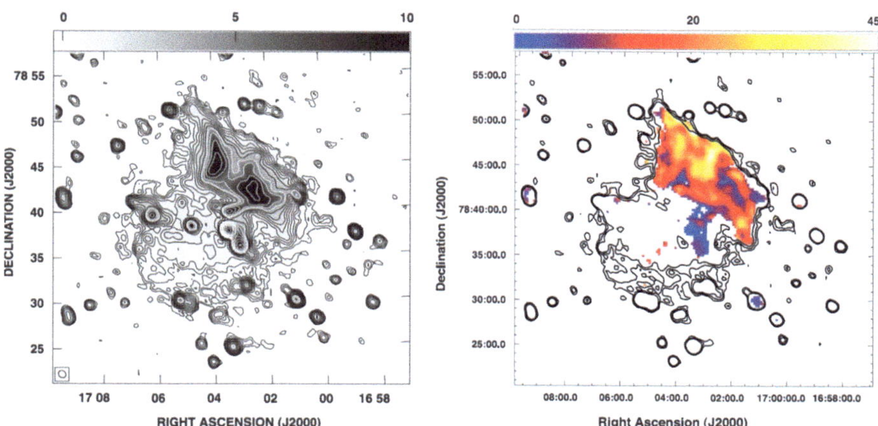

Fig. 3.14 The galaxy cluster A 2256 at 20 cm wavelength, with the total (synchrotron) radiation shown in the *left panel*, and the degree of polarisation in the *right* one. The peripheral relic structure is strongly polarised, while the central halo is not (Clarke and Ensslin, 2006)

$$= \frac{\sin 2\,\lambda^2\,\Phi_0 - i\,\cos 2\,\lambda^2\,\Phi_0 + i}{2\,\lambda^2\,\Phi_0}$$

$$= \frac{\sin \lambda^2 \Phi_0}{\lambda^2 \Phi_0} \cdot e^{i2\psi}.$$

The bandwidth depolarisation is hence given by

$$p(\lambda) = p(0) \cdot \frac{\sin \Delta\psi}{\Delta\psi}, \qquad (3.21)$$

where (Sect. 3.3.2)

$$\Delta\psi = -2\,\lambda_0^2\,RM \cdot \frac{\Delta\nu}{\nu_0}.$$

At low radio frequencies, bandwidth depolarisation can be very strong, The only way to avoid it then is to split the band into many sub-bands and applying the RM synthesis technique.

3.3.8   RM Synthesis

Doing the Fourier transform of the observed $P(\lambda^2)$ has been impossible until recently. The main reason was the scarce coverage of λ^2-space, consisting of a rather limited number of narrow-band observations at discrete frequencies. This has changed, however, since in particular at low frequencies radio-astronomical receivers and backends have been developed with large relative $(\Delta v / v)$ bandwidths and quasi-spectral modes, facilitating a reasonable coverage and sampling of λ^2-space. Given that one can sample the λ^2 space with a decent coverage by observing at various wavelengths or by splitting the wavebands into individual channels one can conveniently exploit the relation

$$P(\lambda^2) = \int_{-\infty}^{+\infty} F(\Phi) \cdot e^{2i\,\Phi\lambda^2}\,d\Phi.$$

This equation can be inverted if one multiplies its right-hand side by a weight function $W(\lambda^2)$ that vanishes for $\lambda^2 < 0$ and at all wavelengths at which data are missing. Its inversion then is

$$\tilde{F}(\Phi) = F(\Phi) * R(\Phi)$$

$$= K \cdot \int_{-\infty}^{+\infty} \tilde{P}(\lambda^2) \cdot e^{-2i\,\Phi\lambda^2}\,d\lambda^2,$$

where

$$R(\Phi) = K \cdot \int_{-\infty}^{+\infty} W(\lambda^2) \cdot e^{-2i\,\Phi\lambda^2}\,d\lambda^2,$$

and

$$K = \frac{1}{\int_{-\infty}^{+\infty} W(\lambda^2)\,d\lambda^2}.$$

In the above, use has been made of the convolution theorem. The weight function accounts for the limited wavelength coverage. Hence, the original Faraday dispersion function $F(\Phi)$ that we are after is convolved with the Fourier transform $R(\phi)$ of the weight function, thus yielding the modified dispersion function $\tilde{F}(\Phi)$. The tilde denotes the reconstructed quantities, and therefore $\tilde{F}(\Phi)$ is an approximate reconstruction of $F(\Phi)$.

The convolution in Faraday depth space corresponds to a Fourier filter in λ^2-space. The quality of the reconstruction depends primarily on the behaviour of $W(\lambda^2)$. Gaps in λ^2-space increase the 'diffraction effects' in Φ-space. We call $R(\Phi)$ the rotation-measure transfer function (RMTF). The qualitative behaviour of $R(\Phi)$ is shown in Fig. 3.15. A thorough mathematical treatment has been presented by Brentjens and de Bruyn (2005). The technique of RM synthesis delivers images of

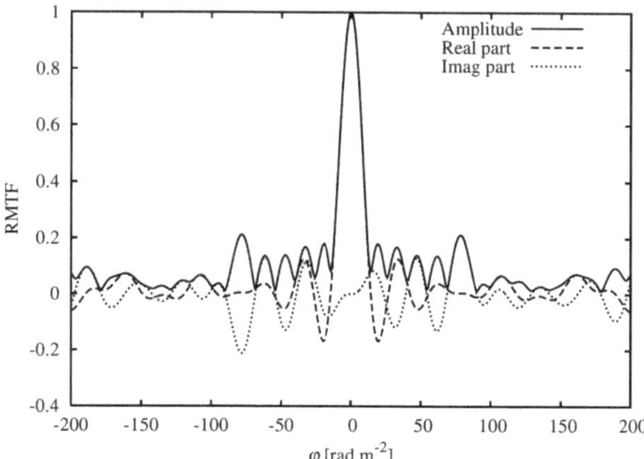

Fig. 3.15 Rotation-measure transfer function (from Brentjens and de Bruyn, 2005)

$\tilde{F}(\Phi)$, i.e. one can evaluate the Faraday dispersion function as a function of Faraday depth Φ, and thus 'penetrate' the magneto-ionised medium. The reconstruction can be greatly improved by deconvolution, using the measured RMTF. This is similar to the clean algorithms used in radio-interferometry, leading to a strong reduction or removal of the sidelobes (here: in Faraday-depth space). At the time of writing this textbook, RM synthesis is in its infancy and provides the most powerful tool for exploiting polarisation measurements with the Westerbork Synthesis Radio Telescope (WSRT), the Karl G. Jansky Very Large Array (VLA), and the newly deployed Low-Frequency Array (LOFAR).

3.4 Zeeman Effect

One of the few tools in astrophysics to directly measure the strength and orientation of magnetic fields is the measurement of the Zeeman effect. Because of the tiny frequency splitting, it 'probes' the densest regions of the ISM, viz. the dense clouds, in particular molecular clouds. Flux freezing (Sect. 4.2.2) implies that the magnetic-field strength is high enough in these regions for the Zeemann effect to become observable.

Its measurement requires the precise determination of the left- and right-circularly polarised components of the radio spectral line. Once measured, it permits detailed studies of the magnetic field in molecular clouds. This kind of information is important for our understanding of star formation out of molecular clouds, in which magnetic fields may play an important role in terms of transporting the angular momentum outwards, which is crucial for any cloud collapse and subsequent star formation to occur. As an example, the Zeeman splitting of the HI line at a frequency of $\nu_0 = 1420.406\,\mathrm{MHz}$ can be measured, delivering three components called δ_1, representing the left-handed elliptical polarisation,

π, representing the linear polarisation, and δ_2, which represents the right-handed elliptical polarisation. If we denote the total (unpolarised) intensity of the radiation as I, then one obtains the following set of measurements:

Frequency	I_r	I_l	Component
$\nu_0 - \Delta\nu$	$I \cdot \frac{(1-\cos\theta)^2}{8}$	$I \cdot \frac{(1+\cos\theta)^2}{8}$	δ_1 (LH ellipt.)
ν_0	$I \cdot \frac{\sin^2\theta}{4}$	$I \cdot \frac{\sin^2\theta}{4}$	π (linear)
$\nu_0 + \Delta\nu$	$I \cdot \frac{(1+\cos\theta)^2}{8}$	$I \cdot \frac{(1-\cos\theta)^2}{8}$	δ_2 (RH ellipt.)

Here, θ is the angle between the line-of-sight and the magnetic field $\vec{B}$, and the shift of the circularly polarised lines is

$$\Delta\nu = \frac{e\,B}{4\,\pi\,m_0\,c} \, ,$$

where m_0 is the electron mass, e the unit charge, and c the speed of light. Plugging in numbers, the total splitting (for the neutral-hydrogen line) is

$$\boxed{\frac{\Delta\nu}{B} = 2.8 \cdot \cos\theta \ \frac{\mathrm{Hz}}{\mu\mathrm{G}} \, .} \qquad (3.22)$$

This is the *Larmor frequency* with which the magnetic moment of the electron precesses in the external magnetic field. It corresponds to the total splitting between the δ_1 and the δ_2 component. The above equation tells us that the Zeeman splitting is extremely small! The typical strength of magnetic fields in the ISM or in molecular clouds is

$$B \approx 10\,\mu\mathrm{G} \ \dots \ 1\,\mathrm{mG}.$$

which gives rise to a frequency splitting of

$$\Delta\nu \approx 28\,\mathrm{Hz} \ \dots \ 2.8\,\mathrm{kHz} \, .$$

This splitting is much smaller than the observed line widths, which are broadened by pressure and turbulence (s.b.). The only way of obtaining a useful Zeeman measurement is the differential measurement of I_r and I_l (see Fig. 3.16):

$$\Delta I(\nu) = I_r(\nu) - I_l(\nu) \, .$$

Figure 3.16 shows that

$$\Delta I = \frac{dI_\nu}{d\nu} \cdot \Delta\nu \, ,$$

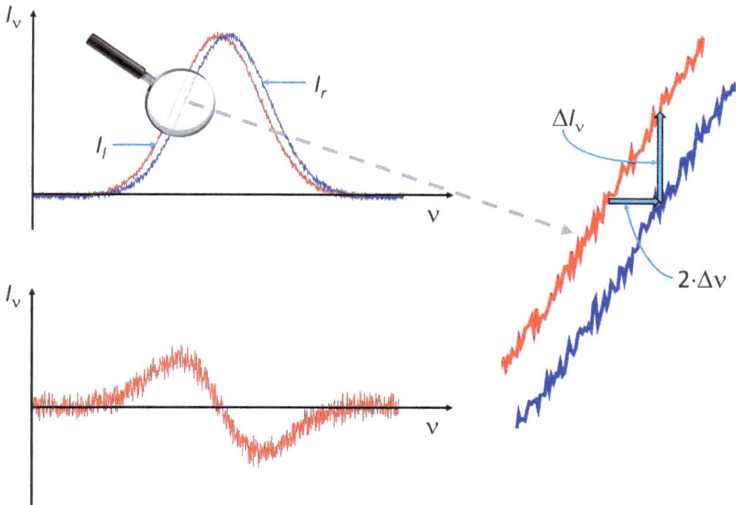

Fig. 3.16 Illustration of the measurement of Zeeman splitting

where $\Delta\nu$ is given by (3.22). Inserting (3.22) we obtain

$$\Delta I(\nu) = 2.8 \cdot B \cos\theta \cdot \frac{dI(\nu)}{d\nu} \,. \qquad (3.23)$$

Here, the magnetic-field strength B is taken in μG, and $d\nu$ in kHz. It should be pointed out that the shape of the differential Zeeman curve determines the direction of $\vec{B}$, i.e. we can figure out whether it points towards or away from us, since ΔI changes its sign.

It is obvious that, owing to the very small splitting of the δ_1 and δ_2 components, radio-astronomical Zeeman measurements are extremely difficult and hence require utterly stable equipment. In Fig. 3.17 a 'good' example of such a measurement in a molecular cloud is displayed. Note how feeble the circularly polarised components are compared to the total intensity of the spectral line. In most cases the Zeeman signatures are much less pronounced than in the example shown here, also because the spectral lines often consist of several components. In this case, the resulting difference between the LHC and RHC components of the line takes a complicated shape. The main difficulty is to measure the small splitting in the presence of Doppler broadening of the emission lines, which is ubiquitous. Expressed in terms of velocity, the Doppler broadening is

$$\Delta v_D = \sqrt{8 \ln 2 \cdot \frac{k\, T_g}{m_H}}\,,$$

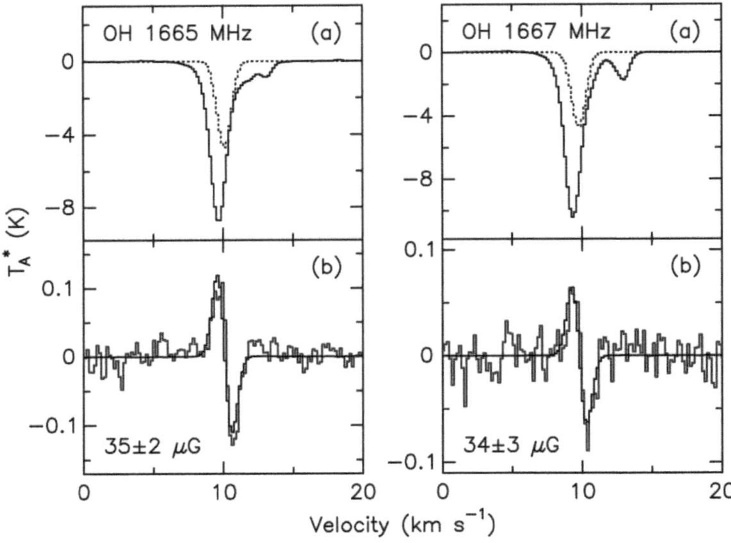

Fig. 3.17 Zeeman effect of the OH absorption line toward Orion B (Bourke et al., 2001)

which via

$$\Delta v_D = \Delta v_D \cdot \frac{v_0}{c}$$

translates into a frequency broadening of (e.g.) the HI line of $\Delta v_D \approx 10\,\text{kHz}$ for a gas temperature of $T_g = 100\,\text{K}$. This obviously exceeds the Zeeman splitting by a large amount.

3.5 Polarised Emission from Dust and Molecules

The alignment of dust grains in the Galactic magnetic field naturally also leads to polarisation of their continuum radiation, observed in the mm, submm, or far-infrared regime. The process that aligns the dust particles has been discussed in Sect. 3.1. Contrary to the optical polarisation, the polarised continuum from dust should have $\vec{E} \perp \vec{B}$, since the electric field emitted as dipole radiation from elongated dust grains has $\vec{E}$ primarily along the long axes of the grains. The expected degrees of polarisation are of order $\leq 10\,\%$. For a seminal article on this topic see Hildebrand (1988).

Weak linear polarisation in rotational emission lines of molecules was first predicted by Goldreich and Kylafis (1982), with the mathematical details being rather complicated, as the strength of the effect depends on the ratio of the cloud's

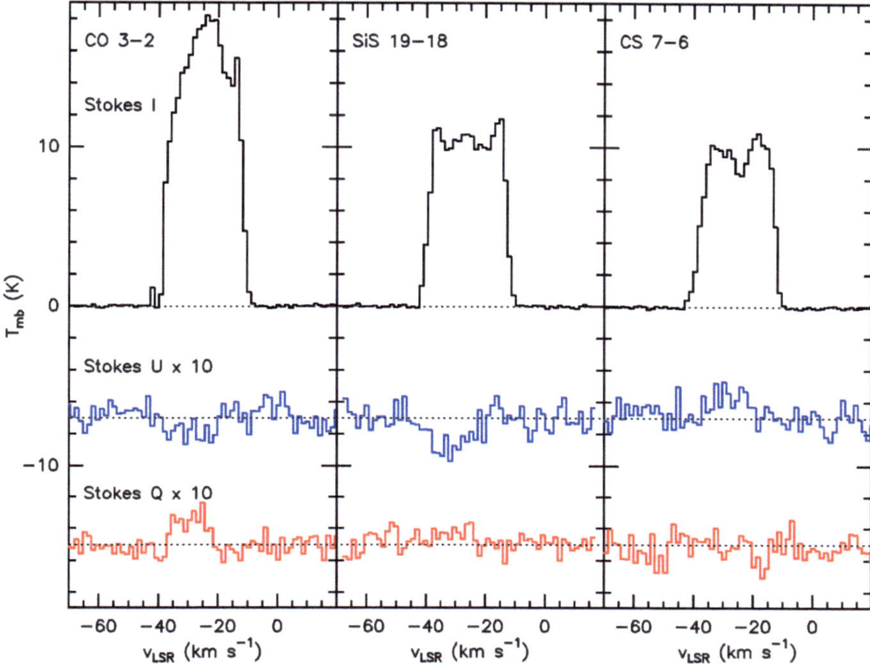

Fig. 3.18 Spectropolarimetry of the CS $J = 2 \rightarrow 1$, SiS $J = 19 \rightarrow 18$, and CO $J = 7 \rightarrow 6$ line towards the evolved star IRC+10216 (Girart et al., 2012)

density compared to the critical density of the transition, on the optical depth of the line, and on the angle between the line of sight and the magnetic field. A weak magnetic field causes the rotational levels to split into magnetic sublevels, and unequal populations of the different sublevels lead to partial linear polarisation. This happens if there is a gradient in the optical depth of a line transition, which may be caused by a velocity gradient or by an anisotropic radiation field.

However, this so-called Goldreich-Kylafis effect is hard to detect, with the degree of linear polarisation just above the artificial instrumental polarisation. Figure 3.18 demonstrates how difficult it is to achieve the required accuracy. Shown is the observed linear polarisation observed using the CS, SiS and CO molecules present in the envelope of the evolved star IRC+10216. The derived degrees of linear polarisation are low, viz. $\approx 2\,\%$ for the CO and SiS lines, and $\approx 4\,\%$ for the CS line. Still today, the detection of the Goldreich-Kylafis effect is a challenge, with very few successful observations reported thus far.

References

Beck, R., 2007, Astron. Astroph. **470**, 539
Bourke, T.L., Myers, P.C., Robinson, G., Hyland, A.R., 2001, Astroph. J. **554**, 916
Brentjens, M.A., de Bruyn, A.G., 2005, Astron. Astroph. **441**, 1217
Burn, B.J., 1966, MNRAS **133**, 67, 1966
Clarke, T.E., Ensslin, T.A., 2006, Astron. J. **131**, 2900
Davis, L., Greenstein, J.L., 1951, Astroph. J. **114**, 206
Gardner, F.F., Whiteoak, J.B., 1966, ARA&A **4**, 245
Girart, J.M., Patel, N., Vlemmings, W.H.T., Rao, R., 2012, Astroph. J. **751**, 20
Goldreich, P., Kylafis, N.D., 1982, Astroph. J. **253**, 606
Hildebrand, R.H., 1988, Quart. J. Astron. Soc. **29**, 327
Scarrott, S.M., Ward-Thompson, D., Warren-Smith, R.F., 1987, MNRAS **224**, 299

Chapter 4
Magnetic Fields Within the Milky Way

4.1 Diffuse ISM

Polarised radio emission from the Milky Way was first detected by Westerhout et al. (1962) and by Wielebinski et al. (1962). This was the final proof that the nonthermal component with its power-law spectrum is synchrotron radiation (free-free emission preponderates in the Galatic plane and in HII regions), and that a magnetic field must be ubiquitously present in the Milky Way. It has been noted early-on that there is little correspondence between the polarised and the total intensity of the synchrotron radiation (see Fig. 4.1), the reason being – as we now know – depolarisation effects, mainly Faraday dispersion (Sect. 3.3.2). The magnetic field is highly tangled: its ordered component produces polarised emission within the radio beam, while its random component, together with the ordered one, gives rise to the total synchrotron intensity. This leads to beam depolarisation, since due to this process the polarised intensity is always lower than the maximum that would be possible. As a consequence, the intensity of the polarised component does not exhibit the same increase towards the ridge of the Galactic plane. Another obvious difference between the two radio maps in Fig. 4.1 is the presence of copious (mostly background) radio sources seen in total intensity, while the polarised intensity one can spot but a few of them. Also here, the reason is depolarisation within these sources, similar to what we witness in our Galaxy nearby. The total synchrotron emission is produced by the total magnetic-field strength

$$B_t = \sqrt{B_u^2 + B_r^2},$$

and therefore has a rather smooth appearance, while the polarised intensity exhibits a rather structured picture. Faraday dispersion and beam depolarisation therefore also limit the 'polarisation horizon' very strongly: while in total intensity one measures the synchrotron radiation all along the line-of-sight through the Milky Way, the

© Springer International Publishing Switzerland 2015
U. Klein, A. Fletcher, *Galactic and Intergalactic Magnetic Fields*,
Springer Praxis Books, DOI 10.1007/978-3-319-08942-3_4

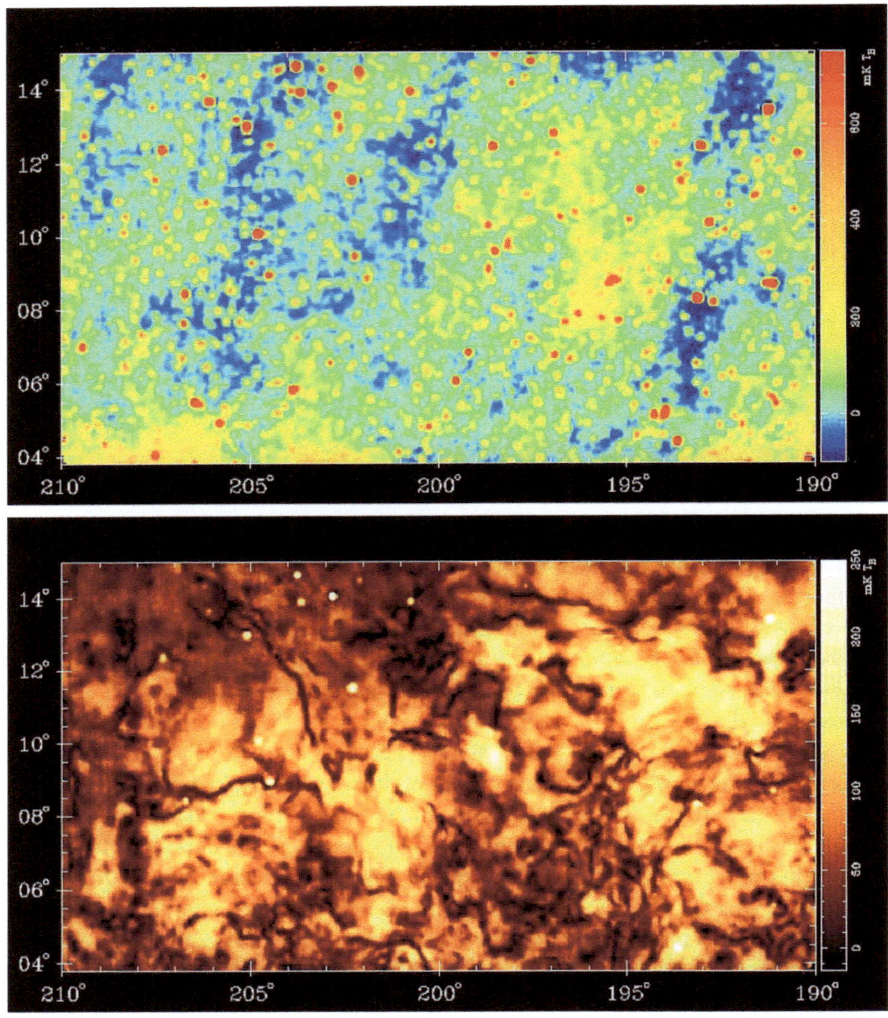

Fig. 4.1 Radio emission from the Milky Way at 1420 MHz, with total intensity in the *top panel*, and polarised intensity in the *lower one* (Uyaniker et al., 1999). Note the different appearances!

distance out to which we see ordered magnetic fields is rather limited, especially within the Galatic plane. Polarisation maps of the Milky Way exhibit conspicuous 'canals', i.e. thread-like structures that are strongly depolarised (Fig. 4.1). These are regions of magnetic-field reversals that lead to counter-Faraday rotation within the beam, and/or regions in which we pick up polarised radiation from orthogonal orientations of the magnetic field within the beam: both effects lead to strong depolarisation (see Sect. 3.3.2).

The diffuse synchrotron radiation is difficult to use as a tool to derive the strength of the magnetic field. While its determination from the total intensity requires the knowledge of the thickness of the source (Sect. 3.2), which is not known for the diffuse medium, the polarised intensity has a very 'limited horizon', as stated above.

A better and more reliable method to determine the magnetic-field strength is the Zeemann effect. From measurements of diffuse HI clouds, a value of $\langle B \rangle \approx 2 \ldots 10\,\mu G$ was obtained. Rotation-measure data from pulsar measurements yield $\langle B \rangle = 2.2 \pm 0.4\,\mu G$. Here, one utilises both, the rotation measure RM as well as the dispersion measure DM, such as to extract the mean magnetic-field strength, weighted by the number density of the thermal electrons

$$\frac{RM}{DM} = K \cdot \frac{\int_0^{r_0} n_e\, B_\parallel\, dr}{\int_0^{r_0} n_e\, dr} \,,$$

so that

$$\langle B_\parallel \rangle = 1.232 \cdot \frac{RM}{DM}\,\mu G \,.$$

The dispersion measure is retrieved from the dispersion of the pulses received from the pulsars, caused by the lag in the refractive ionised medium between the pulsar and the observer. In Fig. 4.2 the distribution of rotation measures from pulsars is shown. It should be mentioned here that minimum-energy or equipartion values derived for external galaxies are in the range $B = 10 \pm 4\,\mu G$ (see Sect. 5.2), so absolutely in line with the above values.

Studying the rotation measure of extragalactic sources probes the whole line-of-sight through the Milky Way, thus facilitating an assessment of the overall morphology of the magnetic field. This can then be compared to models, viz. of the galactic dynamo (see Chap. 6). There have been two studies towards this goal:

- A collection of RMs from all kinds of measurements of extragalactic radio sources in dedicated studies, going back to Simard-Normandin and Kronberg (1980). This has resulted in a couple of thousand of RMs collected over some decades.
- An exploitation of the NVSS (NRAO VLA Sky Survey) at 20 cm wavelength. This has resulted in 37543 RMs, measured at 1435 and 1365 MHz, and has been published by Taylor et al. (2009). An improved version was constructed using a new algorithm by Oppermann et al. (2012, see Fig. 4.3).

From the Galactic distribution of rotation measures, two important pieces of knowledge about the Galactic magnetic field have emerged. First, there is a local vertical field, oriented perpendicular to the Galactic plane, with a field strength of $B_z = 0.3\,\mu G$ below the Galactic plane ($b_{II} < 0$) and $B_z = -0.14\,\mu G$ above the plane ($b_{II} > 0$). This reversal in the sign of B_z across the Galactic

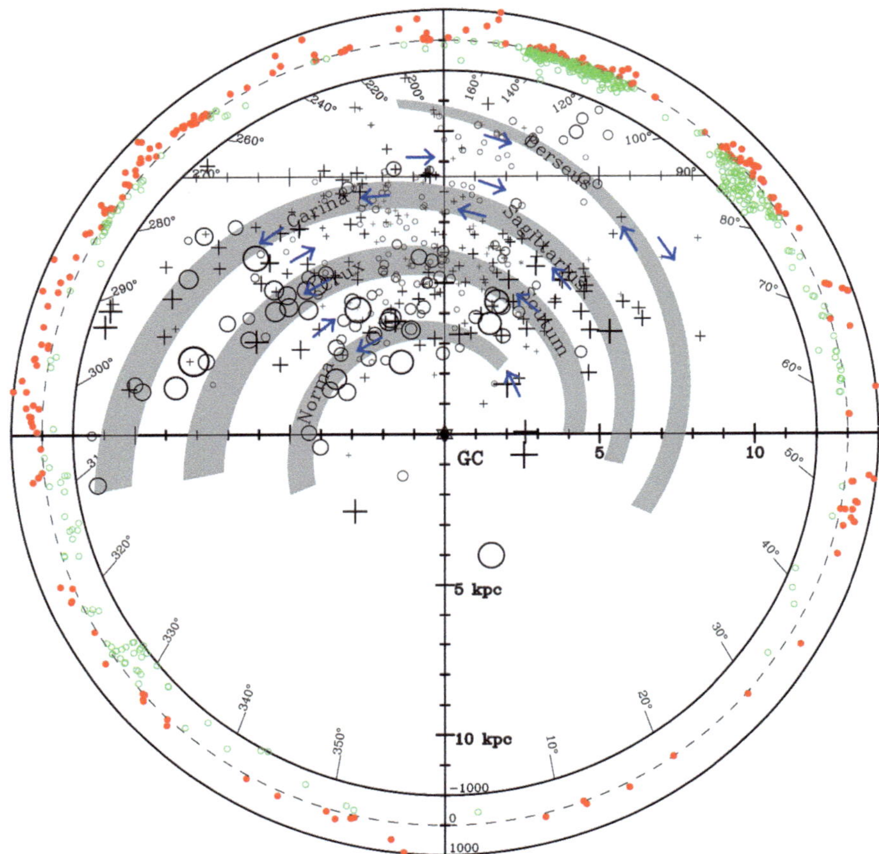

Fig. 4.2 Rotation measures obtained from pulsars with known distances, and a sketch of the Milky-Way spiral arms (From Han, 2007)

mid-plane is consistent with the favoured quadrupolar regular magnetic field pattern of mean-field dynamo theory (see Sect. 5.2.3).

The rotation measures of extragalactic sources suggest an overall axi-symmetric configuration of the magnetic field in the Galactic disk (see Sect. 5.2 for a discussion of the magnetic-field configuration in galaxies). We shall come back to these modes when discussing the observed rotation measures in external galaxies, which provide the more convenient external view. The current state of knowledge has been reviewed by Beck and Wielebinski (2013), who point out that constructing a complete model is not possible at this stage. The current knowledge can be summarised as follows:

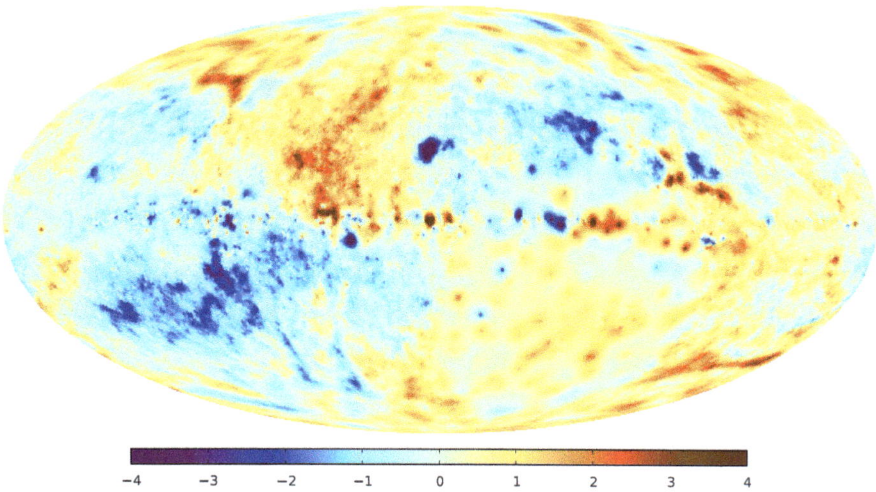

Fig. 4.3 The Galactic Faraday sky (Oppermann et al., 2012)

- Magnetic fields are present everywhere in the Milky Way.
- A large-scale magnetic field, directed clockwise, exists in the Perseus spiral arm.
- A field reversal is evident in the Sagittarius spiral arm.
- The field strengths are $B_u \approx 5\,\mu G$ for the uniform field component and $B_r \gtrsim 5\,\mu G$ for the random one.
- The field strength drops as a function of galacto-centric distance, from $B_t \approx 10\,\mu G$ at $R = 4\,\mathrm{kpc}$, to $B_t \approx 4\,\mu G$ at $R \geq 15\,\mathrm{kpc}$.

Finally, it should be pointed out that the turbulent magneto-ionic medium can best be manifested by juxtaposing the radio continuum maps of the total and polarised radiation using interferometric data. An interferometer acts as a spatial filter that suppresses large-scale structures.[1] Hence, no large-scale structure is seen in total-intensity images of the Galactic synchrotron radiation, whereas images of the polarised intensity exhibit copious small-scale structures. Figure 4.4 demonstrates this impressively: except for some structure in the Galactic plane, the distribution of total intensity is essentially featureless (the image 'only' shows thousands of discrete mostly extragalactic – sources), while the polarised intensity discloses copious plume-like structures emerging from the plane (note, however, that the regular stripes are artefacts). Apart from some source excess along the ridge of the Galactic plane, the total-intensity map also demonstrates the near isotropy in the distribution of distant radio sources as already addressed in Sect. 2.3.

---

[1] See textbooks on Radio astronomy.

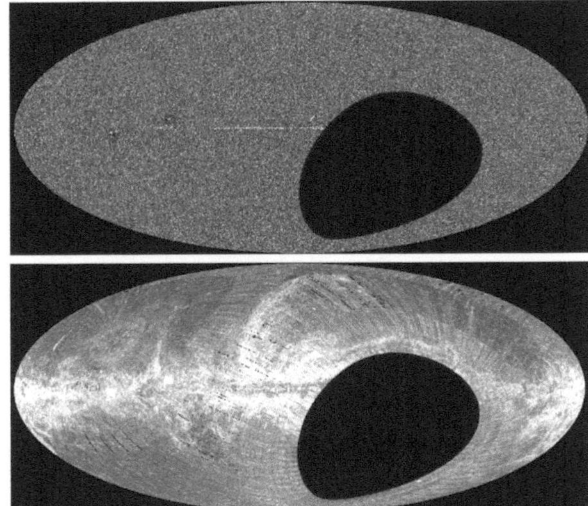

Fig. 4.4 All-sky maps at 1.4 GHz from the NRAO VLA Sky Survey (Rudnick and Brown, 2009), with total (*top*) and polarised intensity (*bottom*). Note the almost isotropic distribution of radio sources in the total-intensity map, which is otherwise featureless, as opposed to the copious structures seen in polarised intensity

4.2   Molecular Clouds and Star-Forming Regions

In order for stars to form, molecular clouds have to contract, which they can do under their own gravity if there is a sufficient critical mass within a certain volume (Jeans mass). Once they collapse, they must rid significant angular momentum to contract to a condensed pre-stellar object.[2] These aspects raise two interesting questions:

- Which process impedes the collapse of all molecular clouds on the scale of a Hubble time?
- Why can stars form at all, given the conservation of angular momentum?

4.2.1   Cloud Stability

It is readily shown that the internal thermal pressure and the rotation of molecular clouds cannot impede their collapse. If we compute the ratio of the internal thermal to the potential energy, we obtain

$$\frac{E_{th}}{|E_{pot}|} \approx \frac{M\,k\,T}{\mu\,m_H} \cdot \left(\frac{G\,M^2}{R}\right)^{-1} = 3 \cdot 10^{-3} \cdot \left(\frac{M}{10^5\,\mathrm{M_\odot}}\right)^{-1} \cdot \left(\frac{R}{25\,\mathrm{pc}}\right) \cdot \left(\frac{T}{15\,\mathrm{K}}\right).$$

---

[2]For a full treatment of this topic see the textbook by Stahler and Palla (2004).

Here, k is the Boltzmann constant and μ is the molecular weight. In case of primordial gas, we have neutral hydrogen plus helium and traces of higher elements, and the molecular weight is $\mu = 1.3$. In molecular clouds having a solar composition, the molecular weight is $\mu = 2.4$. Molecular clouds hardly exhibit any signs of systematic rotation. Typical values deduced for dense cores suggest that clouds are not stabilised against collapse by rotation:

$$\frac{E_{rot}}{|E_{pot}|} \approx \frac{\Omega^2 L^3}{24\, G\, M} = 1 \cdot 10^{-3} \cdot \left(\frac{\Omega}{\text{km s}^{-1}\,\text{pc}^{-1}}\right)^2 \cdot \left(\frac{L}{0.1\,\text{pc}}\right)^3 \cdot \left(\frac{M}{10\,\text{M}_\odot}\right)^{-1}.$$

Here L is the diameter of the clouds, and Ω their angular velocity. Comparing the potential energy with the magnetic energy, the result is drastically different:

$$\frac{E_{mag}}{|E_{pot}|} \approx \frac{B^2\, R^3}{6\,\pi} \cdot \left(\frac{G\, M^2}{R}\right)^{-1} = 0.3 \cdot \left(\frac{B}{20\,\mu\text{G}}\right)^2 \cdot \left(\frac{R}{25\,\text{pc}}\right)^4 \cdot \left(\frac{M}{10^5\,\text{M}_\odot}\right)^{-2}.$$

The fact that magnetic fields can support a cloud against gravitational collapse is the result of 'flux freezing' (Sect. 4.2.2). Typical field strengths derived from Zeeman measurements support this picture: denser clouds or their cores possess stronger magnetic fields, reaching values of mG there. In order to trace the magnetic field deep in a molecular cloud, one needs to observe molecular species that probe high gas densities, such as the CN molecule, which allows to trace magnetic fields at densities $> 10^4$ cm^{-3}. Such measurements had to await radio telescopes operating at mm wavelengths, as the spectral line of CN is observed at $\lambda \approx 3$ mm. In Table 4.1, a small compilation of such Zeeman measurements is given, and Fig. 4.5 shows two examples.

Of course, flux freezing does not imply that the magnetic field is rigidly tied to the molecular clouds, otherwise they would not be able to contract at all. Through a process called 'ambipolar diffusion' (Sect. 4.2.3) the magnetic field manages to 'slip through the gas'. It is important to note that the random component of the interstellar magnetic field also plays a role here: if the magnetic field were entirely uniform, then the gas would simply slip along the field lines, leading to a flat ('pancake'-like) morphology upon contraction. However, this is never observed. The reason must be

Table 4.1 Magnetic-field strengths inferred from Zeeman absorption measurements in Galactic clouds (see also Fig. 4.5)

Object	Type	Diagn.	$B_{\parallel}\,[\mu G]$
Ursa major	Diffuse cloud	HI	+10
L204	Dark cloud	HI	+4
NGC 2024	GMC clump	OH	+87
Barnard 1	Dense core	OH	−27
S 106	HII region	OH	+200
Sgr A/West	Molecular disk	HI	−3000
W75 N	Maser	OH	+3000
DR 21	HII region	CN	−360

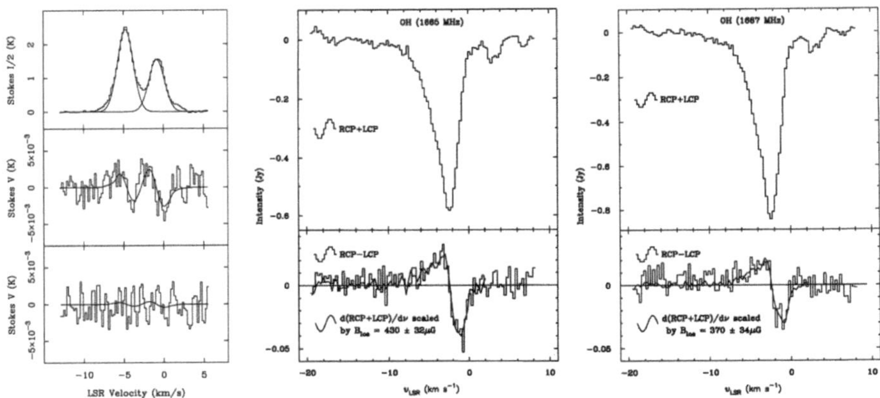

Fig. 4.5 Zeeman measurements in DR 21 (CN, *left*; from Crutcher et al., 1999), and in S 106 (OH line, *right*; from Roberts et al., 1995)

the random component, which 'roughens' the magnetic field, thus preventing such an anisotropic contraction.

4.2.2   Flux Freezing

The coupling of the magnetic field to the ionised gas components is an important process inside and outside condensed astrophysical bodies, as it influences the evolution of stars as well as of the interstellar and intergalactic medium. For instance, this flux freezing on the one hand prevents the consumption of gas in star formation on time scales much shorter than the Hubble time, while on the other hand it supports star formation during the final collapse when angular momentum has to be ridded. Since the universe is predominantly a plasma, flux freezing is inevitable in essentially all astrophysical situations and processes.

A magnetic field threading a molecular cloud must be generated by a current density $\vec{j}$, which is related to the magnetic field $\vec{B}$ by Ampere's law:

$$\vec{\nabla} \times \vec{B} = \frac{4\,\pi}{c} \cdot \vec{j}.$$

Here, the displacement current $\frac{1}{c} \cdot \frac{\partial E}{\partial t}$ has been neglected in view of the low frequencies of interest. The current is carried by free electrons and ions (charged dust grains have too low number densities to be significant). Therefore, we can write it in the form

$$\vec{j} = n_i \, e \, \vec{v}_i - n_e \, e \, \vec{v}_e \approx n_e \, e \, (\vec{v}_i - \vec{v}_e). \tag{4.1}$$

Here, it was assumed that singly charged species (Fe^+, Mg^+, ...) preponderate, so that charge neutrality requires $n_i = n_e$. In order to see how the current is generated by the local electric field, we generalise Ohm's law (which holds for a medium at rest) by shifting temporarily to a reference frame that is moving at the local velocity $\vec{v}$ of the neutral matter. Indicating this reference frame by primes, we have

$$\vec{j}' = \sigma \cdot \vec{E}' = \vec{j}, \tag{4.2}$$

where σ is the electric conductivity. The second equality results from (4.1), noting that neither n_e nor $\vec{v}_i - \vec{v}_e$ can change in the new reference frame. It should be pointed out that we have neglected relativistic corrections of order $(v/c)^2$. To the same accuracy, the new electric field is then given by

$$\vec{E}' = \vec{E} + \frac{\vec{v}}{c} \times \vec{B}. \tag{4.3}$$

Equations (4.2) and (4.3) imply that the generalised Ohm's law now reads

$$\vec{j} = \sigma \cdot \left(\vec{E} + \frac{\vec{v}}{c} \times \vec{B} \right). \tag{4.4}$$

We can now insert this current density into Ampere's law to obtain

$$\vec{\nabla} \times \vec{B} = \frac{4\pi\sigma}{c} \cdot \left(\vec{E} + \frac{\vec{v}}{c} \times \vec{B} \right).$$

We can finally eliminate $\vec{E}$ by using Faraday's law,

$$\vec{\nabla} \times \vec{E} = -\frac{1}{c} \cdot \frac{\partial \vec{B}}{\partial t},$$

and obtain

$$\frac{\partial \vec{B}}{\partial t} = \vec{\nabla} \times (\vec{v} \times \vec{B}) - \vec{\nabla} \times \left(\frac{c^2}{4\pi\sigma} \vec{\nabla} \times \vec{B} \right). \tag{4.5}$$

Equation (4.5) is the fundamental magneto-hydrodynamic (MHD) equation for the magnetic field. The second term on the r.h.s. describes the Ohmic dissipation, which vanishes if the conductivity becomes very large. The electric conductivity σ is a measure of the friction of the charged particles in the conductor, i.e. of its resistance. The particles interact via Coulomb collisions, the frequency of which depends on the particles' density and temperature. The conductivity hence also depends on that frequency (Spitzer, 1956):

$$\sigma = \frac{e^2\, n_e}{m_e\, \nu_{coll}}.$$

The MHD equation is also frequently referred to as the induction equation. Recalling that

$$\vec{\nabla} \times (\vec{\nabla} \times \vec{B}) = \vec{\nabla}\,(\vec{\nabla} \cdot \vec{B}) - \Delta\vec{B} = -\Delta\vec{B}$$

and defining the magnetic diffusivity

$$\eta = \frac{c^2}{4\,\pi\,\sigma},$$

we have, for a homogeneous diffusivity,

$$\frac{\partial\vec{B}}{\partial t} = \vec{\nabla} \times (\vec{v} \times \vec{B}) + \eta\,\Delta\vec{B}$$

or, for an inhomogeneous one

$$\frac{\partial\vec{B}}{\partial t} = \vec{\nabla} \times (\vec{v} \times \vec{B}) - \vec{\nabla} \times \eta\,(\vec{\nabla} \times \vec{B}).$$

Assuming that ions and electrons have the same temperature, the resulting conductivity becomes

$$\sigma \approx 18 \cdot \left(\frac{T}{10^4\mathrm{K}}\right)^{\frac{3}{2}} \mathrm{s}^{-1}. \tag{4.6}$$

Let us assume a static case, i.e. $\vec{v} = 0$, and a constant diffusivity. Then

$$\frac{\partial\vec{B}}{\partial t} = \eta\,\Delta\vec{B} \tag{4.7}$$

This equation implies that the magnetic field leaks through the material from one location to the next. If in this process oppositely directed magnetic-field lines come together they neutralise each other. One can use dimensional arguments here to estimate the time scale for the corresponding decay. For an order-of-magnitude calculation, we simply rewrite the differential equation (4.7) in the simple form

$$\frac{B}{\tau_{diff}} = \eta \cdot \frac{B}{L^2},$$

where τ_{diff} is the diffusion time scale and L is a length scale over which the diffusion happens. This leads to an estimate of the diffusion time, which is the time for the magnetic field to decay, i.e.

$$\tau_{\text{diff}} \approx \frac{L^2}{\eta}.$$

For galaxies, this implies

$$\tau_{\text{diff}} \approx 10^{17} \left(\frac{L}{1\,\text{kpc}}\right)^2 \cdot \left(\frac{T}{10^4\,\text{K}}\right)^{\frac{3}{2}} \quad \text{year!}$$

What a large time scale! If this were true, then we would not expect magnetic fields to change in galaxies over *many orders of magnitude of a Hubble time*. The situation changes, however, dramatically if turbulent motions are added to the system. If $\vec{v} \neq 0$ the magnetic field does not at all survive that long. If we apply this to the Sun, where we have $L \sim 1.4 \cdot 10^{11}\,\text{cm}$ and $T \sim 10^7\,\text{K}$ in the corona, we obtain $\tau_{\text{diff}} \approx 10^{10}$ years, while the solar magnetic cycle is a mere 22 year – an obvious contradiction unless there are (turbulent) motions at play.

The evolution of magnetic fields can be calculated now, given any initial configuration once we know η and $\vec{v}$. Neglecting the right-hand term of the hydrodynamic equation (4.5) reduces it to the ideal MHD equation

$$\frac{\partial \vec{B}}{\partial t} = \vec{\nabla} \times (\vec{v} \times \vec{B}),$$

which is the mathematical representation of *flux freezing*. To see why, we imagine a closed loop $\mathcal{C}$ of comoving gas encircling a surface $\mathcal{S}$ (see Fig. 4.6). We use the Lagrangian derivative

$$\frac{D}{Dt} = \frac{\partial}{\partial t} + \vec{v} \cdot \vec{\nabla},$$

which is the comoving rate of change of a quantity in a particular fluid element momentarily located at $\vec{r}$. This operator describes the rate of change at that location

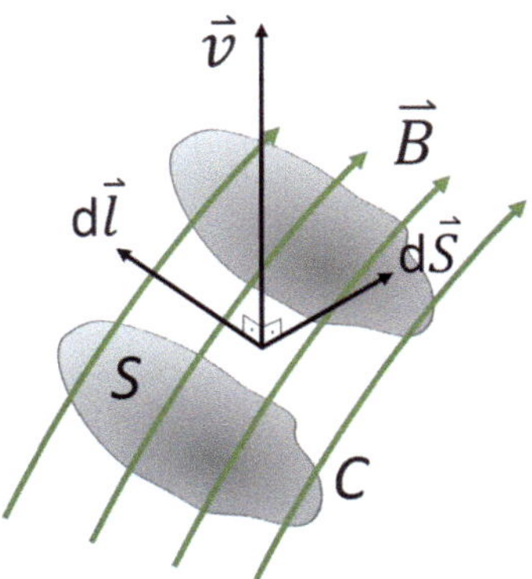

Fig. 4.6 Illustration of flux freezing

$\vec{r}$, plus the spatial derivative in the direction of the fluid velocity multiplied by the magnitude of the fluid velocity. The flux is

$$\Phi = \int_S \vec{B} \cdot d\vec{S},$$

and its rate of change then is

$$\frac{D\Phi}{Dt} = \int_S \frac{\partial \vec{B}}{\partial t} \cdot d\vec{S} + \oint_C (\vec{v} \times d\vec{l}) \cdot \vec{B}.$$

There are two contributions to the change of $\vec{B}$ as a function of time:

- There is a change in the magnetic flux density, due to external causes (e.g. an external force moving the loop), i.e. the change is simply

$$\int_S \frac{\partial \vec{B}}{\partial t} \cdot d\vec{S} \bigg|_{ext}.$$

- The motion of the loop induces an electric field

$$\vec{E} = \frac{\vec{v}}{c} \times \vec{B}. \tag{4.8}$$

Then, because

$$\vec{\nabla} \times \vec{E} = -\frac{1}{c} \cdot \frac{\partial \vec{B}}{\partial t},$$

there will be a contribution

$$\int_S \frac{\partial \vec{B}}{\partial t} \cdot d\vec{S} \bigg|_{move} = -\int_S \vec{\nabla} \times \left(\vec{v} \times \vec{B} \right) \cdot d\vec{S}.$$

Adding these two contributions then yields

$$\frac{D}{Dt} \int_S \vec{B} \cdot d\vec{S} = \int_S \frac{\partial \vec{B}}{\partial t} \cdot d\vec{S} - \int_S \vec{\nabla} \times \left(\vec{v} \times \vec{B} \right) \cdot d\vec{S}$$

$$= \int_S \left[\frac{\partial \vec{B}}{\partial t} - \vec{\nabla} \times \left(\vec{v} \times \vec{B} \right) \right] \cdot d\vec{S} \stackrel{!}{=} 0.$$

So, having used Stokes' theorem (also referred to as Gauss' theorem), we see that the flux through the comoving surface is constant in time.

We can now conceive how magnetic fields help to support molecular clouds to not collapse instantly, i.e. within much less than a dynamical time scale. The question now arises why then stars may form at all from gravitational collapse. The way out is ambipolar diffusion, which describes how the neutral gas can 'slip through' the plasma – which is tightly coupled to the magnetic field – and collapse in a slowed-down mode.

4.2.3   Ambipolar Diffusion

In the last phase of the contraction of a cloud towards a forming star the magnetic field that has helped to get rid of angular momentum must be able to 'slip through' the neutral gas. Otherwise, flux conservation would lead to excessively large field strengths, which would impede the final collapse and which are not observed anyway. This process is referred to as *ambipolar diffusion*. Resisting the relative drift of the neutrals w.r.t. ions, there will be a frictional (drag) force (per unit volume), due to collisions between the ions and neutrals (electrons can be neglected here as they carry much less momentum):

$$\vec{f_d} = \gamma \, \rho_n \, \rho_i \cdot (\vec{v}_i - \vec{v}_n), \tag{4.9}$$

where ρ_n is the mass density of neutrals, ρ_i that of the ions, and γ the drag coefficient. To see how this equation comes about, let us first calculate the collision rate of the ions with any neutral:

$$\nu_{coll,i} = n_i \cdot \langle \omega \, \sigma_{in} \rangle .$$

Here, n_i is the number density of ions,[3] ω is relative velocity of the ions as seen from rest frame of the neutrals, σ_{in} is the cross section for inelastic scattering between the ions and the neutrals, and $\langle \rangle$ stands for averaging over the distribution function of the ions. Denoting with m_i the mass of ions and with m_n the mass of the neutrals, each collision transfers momentum from the ion to neutral that is equal to $\Delta \vec{p} = m_i \cdot (\vec{v}_i - \vec{v}_n) \times$ the fraction of mass of the collision pair contained by the neutral, i.e. $m_n/(m_i + m_n)$. In actual molecular clouds, we have $m_n \approx 2.3 \cdot m_H$ and $m_i \approx 29 \cdot m_H$. We can thus write the momentum $\Delta \vec{p}$ transferred per unit volume and unit time ΔV from the ions to the neutrals as

$$\nu_{coll,i} \cdot \frac{\Delta \vec{p}}{\Delta V} = n_n \cdot \frac{m_n \, m_i}{m_n + m_i} \cdot (\vec{v}_i - \vec{v}_n) \cdot n_i \cdot \langle \omega \, \sigma_{in} \rangle \overset{!}{=} \vec{f_d}$$

with

$$\rho_n = m_n \cdot n_n, \qquad \rho_i = m_i \cdot n_i.$$

---

[3]In dense molecular clouds, HCO^+ is the dominant ion.

We finally recover Eq. (4.9) and realise that

$$\gamma = \frac{\langle \omega \, \sigma_{in} \rangle}{m_n + m_i}.$$

Draine et al. (1983) found

$$\gamma = 3.5 \cdot 10^{13} \, \text{cm}^3 \, \text{g}^{-1} \, \text{s}^{-1}.$$

Ionisation in molecular clouds is caused by cosmic rays. Theoretical calculations show that for densities $n_n \approx 10^4 \, \text{cm}^{-3}$ the fractional ionisation is $n_i/n_n \approx 10^{-7}$, in good agreement with what is derived observationally for molecular clouds (i.e. deriving abundances of neutrals and ions). Now we calculate the drift speed. The force on a charged particle is

$$\vec{F} = q \cdot \vec{E} + \frac{q}{c} \cdot \vec{v} \times \vec{B}.$$

In order to calculate the force on a fluid of charged particles it is convenient to replace charge by charge density and $q \, \vec{v}$ by current density:

$$q \to \sigma, \, q\vec{v} \to \vec{j}.$$

Then the force per unit volume is the Lorentz force[4]

$$\vec{f_L} = \sigma \cdot \vec{E} + \frac{1}{c} \cdot \vec{j} \times \vec{B}$$

and, noting that in general $E^2/B^2 \ll 1$, we have

$$\vec{f_L} = \frac{1}{c} \vec{j} \times \vec{B}.$$

Using Maxwell's equation

$$\vec{\nabla} \times \vec{B} = \frac{4\pi}{c} \cdot \vec{j},$$

we finally derive for the Lorentz force density

$$\vec{f_L} = \frac{1}{4\pi} (\vec{\nabla} \times \vec{B}) \times \vec{B}.$$

---

[4]It is a bit inconvenient to the reader that we are using the symbol σ for the electric conductivity, the charge density, and for cross sections.

The dominating forces on the charged particles are now the drag and the Lorentz force, all others can be neglected. These two forces must add to zero, i.e. $\vec{f}_d = \vec{f}_L$, and hence

$$\gamma\,\rho_n\,\rho_i\cdot(\vec{v}_i - \vec{v}_n) = \frac{1}{4\,\pi}(\vec{\nabla}\times\vec{B})\times\vec{B}.$$

Then, to order of magnitude, the drift velocity $v_d = v_i - v_n$ is

$$v_d \approx \frac{B^2}{4\,\pi\,\gamma\,\rho_n\,\rho_i\,L} \approx \frac{v_A^2}{V},$$

where we have introduced the Alfvén speed

$$v_A = \frac{B}{\sqrt{4\,\pi\,\rho}}$$

by assuming $\rho = \rho_n + \rho_i \approx \rho_n$ for the low ionisation degrees (s.a.). The speed V contains the product

$$V = \gamma\,\rho_i\,L.$$

Inserting typical values, i.e. $L = 0.1\,\text{pc}$ and $n_n = \rho/m_n = 10^4\,\text{cm}^{-3}$, we obtain $V = 6\,\text{km s}^{-1}$. With $B \approx 30\,\mu\text{G}$ we have $v_A \approx 0.4\,\text{km s}^{-1} \ll V$, and the drift speed is a mere $v_d = 27\,\text{m s}^{-1}$. The time scale for ambipolar diffusion then is

$$t_{AD} \approx \frac{L}{v_d} \approx 3.6\cdot 10^6 \text{ year,}$$

using the above quantities. This is an important process, as the time scale t_{AD} is of the same order of magnitude as the free-fall time of the cloud collapse,

$$\tau_{ff} = \left(\frac{3\,\pi}{32\,G\,\rho_n}\right)^{\frac{1}{2}} = 3.1\cdot 10^7\cdot\left(\frac{n_n}{\text{cm}^{-3}}\right)^{-\frac{1}{2}} \text{ year.}$$

Note that the tight coupling of the ions to the magnetic field is hardly disturbed by the collisions, since the collision frequency (or rate) is much lower than the Larmor frequency. Hence, the ions may take many revolutions about the magnetic field before being knocked off by a neutral. In Fig. 4.7 examples of magnetic-field structures observed in molecular clouds are shown. The image in the left panel shows the magnetic-field strength toward a molecular outflow in the HII region S 106, superimposed onto the 18-cm continuum emission. The Zeeman map suggests the magnetic field to be aligned with the major axis of the object as depicted by the continuum emission. In the central panel, the magnetic-field structure in a young stellar object close to the compact HII region DR 21 as

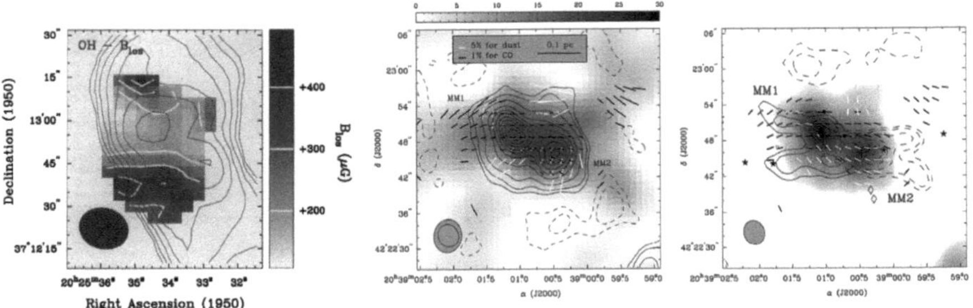

Fig. 4.7 Examples of magnetic-field structures observed in molecular clouds. *Left*: magnetic-field strength in S 106, obtained from OH observations at $\lambda = 18$ cm conducted with the VLA (Roberts et al., 1995). The contours depict the continuum emission. *Centre*: structure of the magnetic field in DR 21, inferred from polarisation measurements of the dust emission at $\lambda = 1.3$ mm and in the ^{12}CO(2-1) line with BIMA (Lai et al., 2003). *Right*: Polarisation vectors, with CO iso-velocity contours superimposed

deduced from dust (white bars) and CO (black bars) polarisation measurements (Goldreich-Kylafis effect) is shown, superimposed onto the ^{12}CO(2-1) intensity (gray scale) and the dust continuum at $\lambda = 1.3$ mm (contours). The right panel presents a superposition of the magnetic-field structure onto CO iso-velocity contours showing the outflow of this object (dashed contours: blue shifted; solid contours: red-shifted). Also here, the magnetic field appears to be parallel to the major axis of this object.

4.3 Supernova Remnants

Among the brightest radio sources in the sky are supernova remnants (SNR). These were also among the first 'discrete sources' discovered at radio frequencies, such as Cas A (or 3C 461), a radio image of which is presented in Fig. 4.8, showing the expanding relic of a supernova explosion that occured[5] in 1680, and Tau A (3C 75, M 1), seen in 1054 by Chinese astronomers. There are basically two types of SNR:

- Plerions ('filled' SNR)
- Shell-type SNR.

From optical spectroscopy and VLBI observations one deduces the initial expansion speeds, reaching $v_{exp} \approx 20\,000\,\mathrm{km\,s^{-1}}$. Their shells initially expand freely, owing to their much larger density as compared to their surroundings. A collisional shock is expected when a shell has swept over distance that is about equal to the mean free path of the protons. A proton moving with a speed of $v = 20\,000\,\mathrm{km\,s^{-1}}$ has a kinetic energy of $E_{kin} \approx 2\,\mathrm{MeV}$. Its mean free path can be calculated as follows. It is given by the path it takes a cosmic-ray proton to lose

---

[5]This means when the light of the event reached the earth.

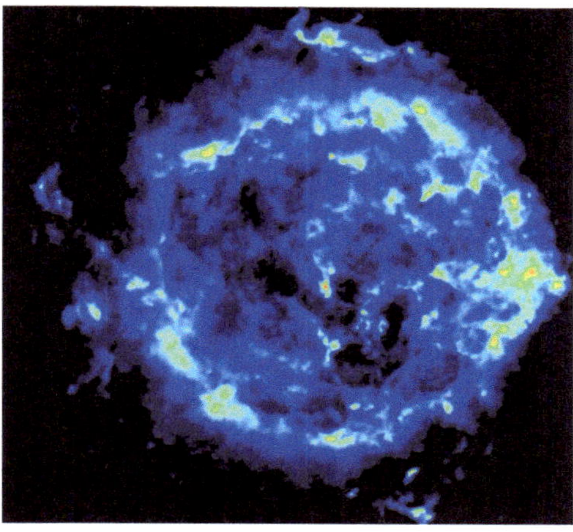

Fig. 4.8 Radio continuum image of the SNR Cas A obtained with the VLA at 5 GHz (Braun et al., 1987)

its kinetic energy via ionisation of the hydrogen atoms or H_2 molecules. The cross section for this is (Bethe, 1930):

$$\sigma_{CR} = \frac{2\pi\, e^4\, Z^2}{m_e\, \chi_0\, v^2} \cdot 0.285 \cdot \left\{ \ln \frac{2\, m_e\, v^2}{\left[1 - \left(\frac{v}{c}\right)^2 \cdot \chi_0\right]} + 3.04 - \left(\frac{v}{c}\right)^2 \right\},$$

where χ_0 the ionisation energy (13.6 eV in our case) and v the speed of the ionising protons. With $v = 2 \cdot 10^9\,\mathrm{cm\,s^{-1}}$ and $Z = 1$ we find $\sigma_{CR} \approx 10^{-17}\,\mathrm{cm^2}$, and with a mean gas density of $\bar{n}_{HI} \approx 1\,\mathrm{cm^{-3}}$ we obtain a path between two subsequent ionisations of $l_i \approx 0.03\,\mathrm{pc}$. Since the proton loses only about 10^{-4} of its kinetic energy in each ionisation, the total mean free path becomes $l = 10^4 \cdot l_i = 300\,\mathrm{pc}$. Obviously, no classical shock would develop under such conditions. However, in the presence of a magnetic field of strength $B \approx 5\,\mu\mathrm{G}$, these protons have gyration radii of (Sect. 2.3.1)

$$r_L = \frac{m\, v\, c}{e\, B} \approx 10^{-8}\,\mathrm{pc}.$$

They are therefore tightly coupled to the magnetic field, which thus gives rise to a massless barrier! The result is a hydromagnetic or MHD shock. The details of scattering in this process are as yet not fully understood. The supernova blast wave will thus plow through the ISM and compress the magnetic field, thereby enhancing its strength within the shell. We schematically distinguish four phases in the evolution of an SNR.[6]

---

[6]Not all SNR fit into this scheme. For instance the Crab Nebula and all plerions possess a morphology that is dominated by a pulsar wind, which transfers rotational energy into the SNR.

4.3.1   *Free Expansion*

At this early stage, the shell's radius behaves like

$$R \propto t,$$

and the mass of the swept-up gas is still less than the initially injected mass M_S of the shell

$$\frac{4}{3}\,\pi\,R_S^3\,\rho_{ext} \leq M_S.$$

Assuming for the surrounding gas density a value of $\rho_{ext} \approx 2 \cdot 10^{-24}\,\mathrm{g\,cm^{-3}}$ and an initial shell mass of $M_S = 0.25\,\mathrm{M_\odot}$, we obtain $R_S \approx 1.3\,\mathrm{pc}$, so this phase will last for about

$$t_1 \approx \frac{R_S}{v_{exp}} \approx 60 \text{ year.}$$

This phase can be observed in extragalactic environments using VLBI-monitoring observations, the spectacular example of SN 1993j shown in Fig. 4.9. Their monochromatic radio luminosities decrease with a power-law as a function of time, with emission appearing at a high frequency first, and showing up at lower frequencies as time elapses, because the medium becomes less opaque.

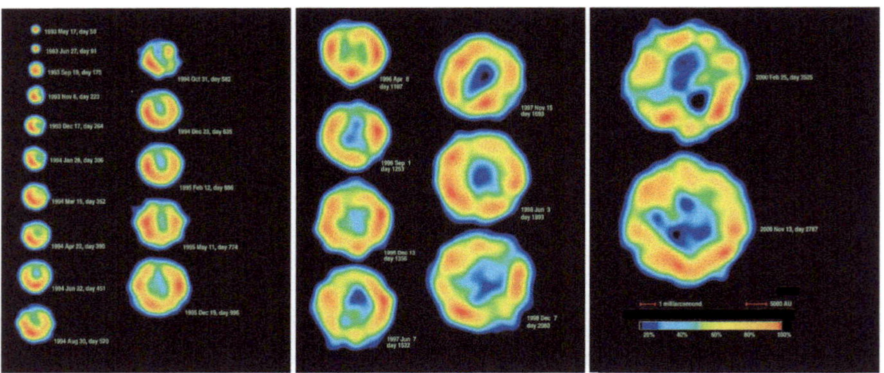

Fig. 4.9 VLBI observations of the supernova SN1993j, showing its expansion (Bietenholz et al., 2003)

4.3.2 Adiabatic Phase

When the swept-up mass becomes greater than the ejected mass, the dynamics is described by the adiabatic similarity solution that goes back to Sedov (1959). At this stage, the temperature of the shocked gas is so high that its radiation is relatively weak. Hence the only important energy loss is through the adiabatic expansion of the gas. This phase terminates when the temperature drops below $T \sim 10^6$ K. During this phase, the radius of the shell behaves as

$$R \propto n^{-\frac{1}{5}} \cdot t^{\frac{2}{5}}.$$

For the case of spherical symmetry, the one-dimensional numerical solution of the hydrodynamic equations has been presented by Chevalier (1974), yielding a relation between the supernova energy E_0, the surrounding gas density n_0, and the expansion speed and the radius, such that

$$E_0 = 5.3 \cdot 10^{43} \cdot \left(\frac{n_0}{\mathrm{cm}^{-3}} \right)^{1.12} \cdot \left(\frac{v}{\mathrm{km\ s}^{-1}} \right)^{1.4} \cdot \left(\frac{R}{\mathrm{pc}} \right)^{3.12} \mathrm{erg}.$$

During this phase the total energy E_0, consisting of thermal energy and mass motion, is constant. This phase is also called Sedov phase, as it goes back to a calculation describing similarities in blast waves (nuclear bombs, supernovae, ...). Since this phase terminates when the temperature drops below $T \sim 10^6$ K (at which point radiative losses become important), its duration can be estimated from

$$T = 1.5 \cdot 10^{11} \cdot \left(\frac{n_H}{\mathrm{cm}^{-3}} \right)^{-\frac{2}{5}} \left(\frac{t}{\mathrm{year}} \right)^{-\frac{6}{5}} \cdot \left(\frac{E}{4 \cdot 10^{50}\ \mathrm{erg}} \right)^{\frac{1}{5}} \mathrm{K}.$$

Inserting $n_H = 1\ \mathrm{cm}^{-3}$ and $E = 4 \cdot 10^{50}$ erg, this yields $t = 2 \cdot 10^4$ year.

4.3.3 Isothermal or Radiative Phase

The main energy loss now is through radiation: at a temperature of $T \leq 10^6$ K, the ions of C, N and O begin to recombine with the free electrons. The collisional excitation of ions so formed and the line radiation they emit increase the cooling rate of the gas by orders of magnitude. Hence, the shock becomes radiative. The radius of the SNR is now $R_S \approx 15$ pc for a density of $n_H = 1\ \mathrm{cm}^{-3}$, the age is $\sim 4 \cdot 10^4$ year, and the expansion speed has dropped to $v_{exp} \approx 85\ \mathrm{km\ s}^{-1}$. Conservation of momentum implies that $M \cdot v_{exp} = $ constant, i.e.

$$\frac{4}{3} \pi R_S^3 \rho_{ext}\, v_s = const,$$

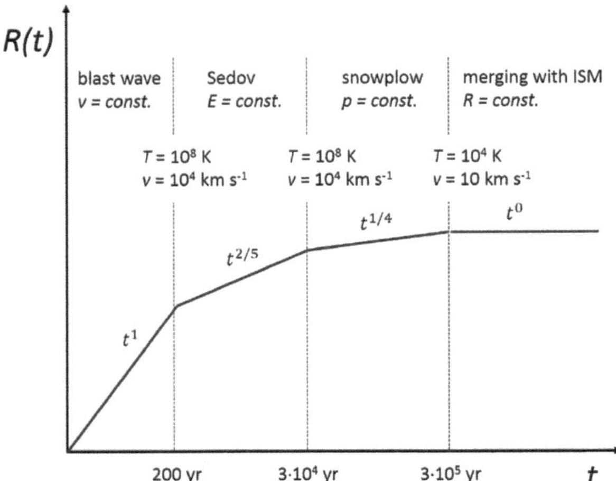

Fig. 4.10 Illustration of the different evolutionary phases of SNR

which by integration yields

$$R_S = R_{rad} \cdot \left(\frac{8}{5} \cdot \frac{t}{t_{rad}} - \frac{3}{5} \right)^{\frac{1}{4}}.$$

Here, R_{rad} and t_{rad} refer to the commencement of this radiative phase. This last phase ends with the dispersion of the shell when the expansion speed drops to $\leq 9\,\mathrm{km\,s^{-1}}$, which is the velocity dispersion of the ISM. Figure 4.10 summarises the four evolutionary phases of SNR.

The morphology of magnetic fields in SNR is kind of a mixed bag, even in shell-type remnants. Some of them exhibit a radial, others a tangential field orientation. However, if one subdivides them into young and old, the former show preferentially radial fields, while the latter have tangential fields. The radial fields are probably caused by Rayleigh-Taylor instabilities in the young, rapidly expanding remnants, as is indicated by MHD-simulations, while the tangential field pattern in old remnants is likely to be the result of field compression in the shock wave.

4.4   Acceleration of Cosmic Rays

Cosmic rays (CR) play an important role in plasma astrophysics. The enrichment of the intra-cluster medium (ICM) and the intergalactic medium (IGM) with heavy elements, and along with this also the magnetisation of the ICM or IGM, must have been furnished very early in the universe via galactic winds (Sect. 5.6) produced by starburst galaxies, so dwarf galaxies in the first place. The origin of cosmic rays is to some extent still a matter of speculation. While particles with energies up to 10^{14} eV may well have been produced by SN shock fronts, those with higher energies ($>10^{14}$ eV) still pose a problem in this respect. Active galactic nuclei (AGN, Chap. 7) are candidates for being factories, while the highest energies are likely to be provided by γ-ray bursters (GRBs). These so-called ultra-high energy cosmic rays (UHECR) have energies of up to $E \sim 10^{21}$ eV, and their sites of origin must be searched for within the so-called GKZ horizon[7] (Greisen, 1966; Zatsepin and Kuz'min 1966).

Cosmic-ray electrons are important for the astrophysicist because they reveal much of the relativistic plasmas on all astrophysical scales by virtue of synchrotron radiation, and we always must keep in mind that each relativistic electron is accompanied by a relativistic proton, with its $\sim$2000 larger rest mass. It is hence justified to at least briefly consider some basics of CR acceleration, while a full treatment would be beyond the scope of this textbook.[8] How much energy can a supernova and its remnant provide to the CR particles? As an example, take Cas A, the brightest radio source in the northern sky. Its flux density is $S_\nu = 2720$ Jy at 1 GHz. With a distance $D = 2.8$ kpc, its monochromatic luminosity is $P_{1\mathrm{GHz}} = 2.6 \cdot 10^{18}$ W Hz^{-1}. Using the minimum energy formula (3.12), we obtain $E_{min} = 2.6 \cdot 10^{48} \, \eta^{\frac{4}{7}}$ erg. If the average time span between subsequent supernova explosions is t_{SN} and the escape time for particles, i.e. the time they need to leave the confinement volume, is t_c, then the mean energy transferred to the cosmic rays in a volume V corresponds to their local energy density u_{CR}, i.e.

$$u_{CR} = \frac{t_C}{t_{SN}} \cdot \frac{E_0}{V} \; \mathrm{erg\, cm}^{-3}.$$

Here, E_0 is the average energy release in high-energy particles per supernova. Suppose that the lifetime of such particles in the Galaxy disk is 10^7 year (Eq. 2.38). With a thickness of 700 pc and a radius of 10 kpc the volume becomes $6 \cdot 10^{66}$ cm^3. The local energy density of high-energy particles is ~ 1 eV cm^{-3}. Adopting a SN rate of about 1 every 30 year, then the average energy release becomes $\sim 3 \cdot 10^{49}$ erg, which should be compared with the energy E_{min} as derived above, using minimum-energy arguments for the relativistic particles and the magnetic field. With $\eta \approx 100$, these energies become comparable!

We now briefly discuss the process of particle acceleration by multiple shocks in the ISM. This mechanism was first treated by E. Fermi (1949) and is referred to as *2nd-order Fermi acceleration*, meaning that it deals with energy gains that are $\propto (v/c)^2$. In Fermi's original picture, charged particles are reflected from 'magnetic mirrors' associated with irregularities in the Galactic magnetic field.[9] The particles may get trapped in a 'magnetic bottle' and are reflected forward and backward between moving clouds, which have a frozen-in magnetic field (Fig. 4.11). It will

---

[7]Particles (protons) with energies as high as this interact with the CMB photons, which they 'see' as γ-rays in their rest frame. The CMB hence becomes opaque to them on a scale of about 50 Mpc. The interaction implies photo-pion production.

[8]In this section, we follow the discussion of Longair (2011).

[9]First-order Fermi acceleration happens when only head-on collisions occur.

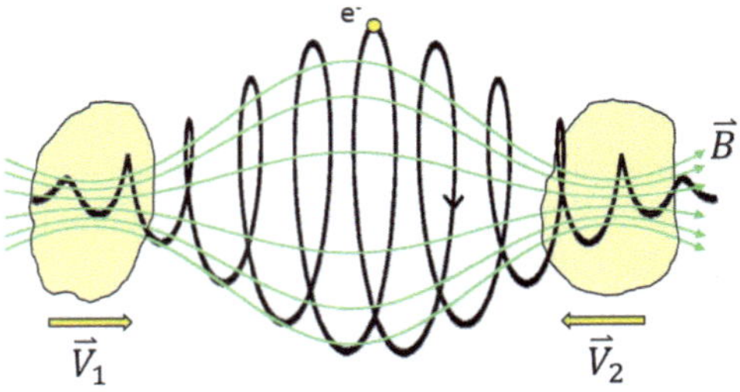

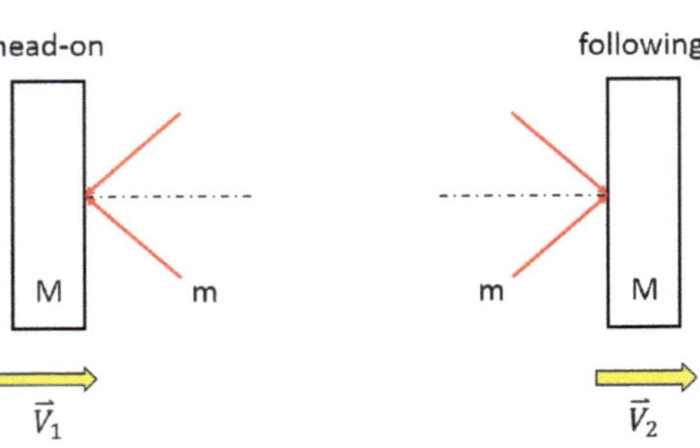

Fig. 4.11 Illustration of particle trapping in a 'magnetic bottle'

Fig. 4.12 Sketch of the geometry in Fermi-II acceleration, with head-on and following collision

be shown here that this process establishes a power-law energy distribution of the accelerated particles, as is observed.

Let us calculate the energy gain of a particle for an angle θ between the initial motion of the particle and the normal to the surface of the mirror surface (the cloud surface, as sketched in Fig. 4.12). It is important here to carry out a proper relativistic analysis. We assume the mass M of the cloud to be infinitely large compared to the particle's mass m, $M \gg m$, so that the cloud's velocity V is unchanged in the collision. The centre-of-momentum frame is therefore that of the cloud. The energy of the particle in this frame is

$$E' = \gamma_V \cdot (E + V\,p\cos\theta),$$

where

$$\gamma_V = \left(1 - \frac{V^2}{c^2}\right)^{-\frac{1}{2}}.$$

is the Lorentz factor referring to the cloud's motion. Here, the primed quantities refer to the cloud's frame, which is also the centre-of-momentum frame. The x-component of the relativistic three-momentum in the centre-of-momentum frame is

$$p'_x = p'\cos\theta = \gamma_V\left(p\cos\theta + \frac{V\,E}{c^2}\right).$$

In the collision, the particle's energy is conserved, i.e. $E'_{before} = E'_{after}$, and its momentum in the x-direction is reversed, i.e. $p'_x \to -p'_x$. Transforming back to the observer's frame, we hence find

$$E'' = \gamma_V \left(E' + V\, p'_x \right).$$

Inserting E' and p'_x from above and recalling that

$$\frac{p_x}{E} = v \cdot \frac{\cos\theta}{c^2}$$

we find

$$E'' = \gamma_V^2\, E \cdot \left[1 + \frac{2\,V\,v\,\cos\theta}{c^2} + \left(\frac{V}{c}\right)^2 \right].$$

Expanding this to second order in $\frac{V}{c}$, we arrive at

$$E'' - E = \Delta E = \left[\frac{2\,V\,v\,\cos\theta}{c^2} + 2\left(\frac{V}{c}\right)^2 \right] \cdot E. \tag{4.10}$$

We now need to integrate over θ. The probabilities of head-on and following collisions are proportional to the relative velocities of approach of the particle and the cloud:

$$P \propto v + V\cos\theta \qquad \text{head on}$$

$$P \propto v - V\cos\theta \qquad \text{following}$$

Since $v \approx c$, the probabilities are proportional to $1 + V/c\,\cos\theta$, where $0 < \theta < \pi$. Recalling that the probability for the angle to lie in the range θ and $\theta + d\theta$ is proportional to $\sin\theta\,d\theta$, we find, setting $x = \cos\theta$ and averaging over all angles in the range 0 to π for the first term of ΔE in (4.10):

$$\left\langle \frac{2\,V\,\cos\theta}{c} \right\rangle = \left(\frac{2V}{c}\right) \frac{\displaystyle\int_{-1}^{+1} x \cdot \left[1 + \left(\frac{V}{c}\right) x \right] dx}{\displaystyle\int_{-1}^{+1} \left[1 + \left(\frac{V}{c}\right) x \right] dx} = \frac{2}{3}\left(\frac{V}{c}\right)^2.$$

Thus, in the relativistic limit the average gain per collision is

$$\left\langle \frac{\Delta E}{E} \right\rangle = \frac{8}{3} \cdot \left(\frac{V}{c}\right)^2.$$

This illustrates the famous result derived by Fermi, namely that the average increase in energy is only *2nd order* in V/c. It is immediately apparent that this leads to an exponential increase in energy since the same fractional increase occurs in each collision. If the mean free path between the 'mirrors' along a magnetic-field line is L, then the time between collisions is $L/(c \cos \phi)$, where ϕ is the pitch angle of the particle with respect to the magnetic-field direction. The average over $\cos \phi$ yields the average time between the collisions, which is just $2 L/c$.[10] The typical rate of energy increase is therefore

$$\frac{dE}{dt} = \frac{4}{3} \cdot \left(\frac{V^2}{c L} \right) \cdot E.$$

The so-called diffusion-loss equation (Sect. 5.6) reads

$$\frac{dN}{dt} = D \, \vec{\nabla}^2 N(E) + \frac{\partial}{\partial E} [b(E) \cdot N(E)] - \frac{N(E)}{\tau_{esc}} + Q(E). \qquad (4.11)$$

This equation describes the rate of change of the particles in a given volume, where the first term on the left-hand side accounts for diffusion, the second one for the energy losses or gains, the third one for the escape of particles from the volume, while the last term is the source term. We are interested here in the steady-state solution, i.e. we set $dN/dt = 0$. We are not interested in diffusion, hence we set $D\vec{\nabla}^2 N = 0$. In our case, the energy loss term becomes a gain term and is given by

$$b(E) = -\frac{dE}{dt} = -a \, E.$$

Hence, the diffusion loss equation becomes

$$-\frac{d}{dE} [a \, E \, N(E)] - \frac{N(E)}{\tau_{esc}} = 0.$$

Differentiating and integrating, we find

$$\frac{dN(E)}{dE} = -\left(1 + \frac{1}{a \cdot \tau_{esc}} \right) \cdot \frac{N(E)}{E} ,$$

from which we hence obtain

$$N(E) \, dE = A \cdot E^{-g} \, dE,$$

---

10 $\left\langle \dfrac{L}{c \cos \phi} \right\rangle = \dfrac{L}{c} \cdot \dfrac{\int d(\cos \phi)}{\int \cos \phi \, d(\cos \phi)} = \dfrac{2L}{c}.$

where

$$g = 1 + \frac{1}{a \cdot \tau_{esc}} = 1 + \frac{\tau_{acc}}{\tau_{esc}}.$$

This is the power-law that is found for the CR electron energy spectrum and that we made use of in Sect. 2.3.2. The quantity a in the above equation can be considered as the inverse acceleration time scale so that the slope g of the particle energy spectrum depends on the relative balance between acceleration and loss time scales. The full and more realistic calculations are a lot more complicated and have to account for the – here neglected – energy losses of the accelerated particles, which would lose much of their acquired energy via ionisation losses in the first place. Furthermore, in the above treatment we have omitted the fact that cloud motions in the ISM have a mere $V/c \approx 10^{-4}$. In a full treatment, one needs to take the energy losses into account and assume supernova shock fronts as the main accelerators (O'Drury, 1983). Nevertheless, the main result, viz. a power-law for the energy spectrum of relativistic electrons results in what is called the theory of diffusive shock acceleration. It provides particle spectra up to energies relevant for the Galatic synchrotron radiation. Note that even though in Fermi-II acceleration the particles lose energy in the following collisions, there is always a net energy gain since the head-on collisions are somewhat more frequent.

4.5 UHECR

The measured CR energy spectrum (Fig. 2.17) exhibits particles with up to energies of $\sim 10^{21}$ eV! Such particles have been coined ultra-high energy cosmic rays (UHECR). Their origin is still unknown. The Larmor radius

$$r_L = \frac{E}{e\,B}$$

of a proton with this energy in a 1-μG magnetic field is about 1 Mpc! Hence, if these particles originated within the Milky Way, their site of origin should be found within an angle

$$\Delta\theta = \frac{L}{r_L},$$

where L is the traversed (curved) path on the way to us. Again, with $B \approx 1\,\mu$G, we should find the responsible particle accelerator within $\sim 10°$ of the direction in which the CR has been detected (with the Auger experiment, see Fig. 4.13). However, the recorded events are distributed more or less isotropically across the sky, with no preference of locations in the Galactic plane. They must therefore be of extragalactic origin, also because there is not likely to be any process within the Milky Way that could provide the required energies.

On the other hand, the particles must stem from a volume within the so-called GKZ limit (Greisen, 1966; Zatsepin and Kuz'min 1966). The reason is that protons

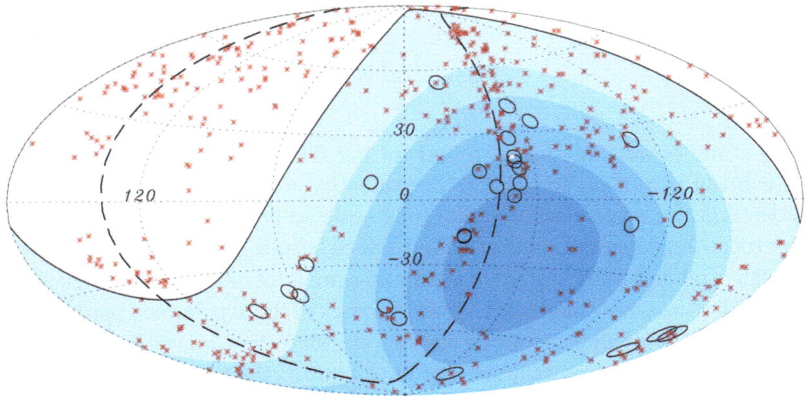

Fig. 4.13 UHECRs recorded with the Auger array

with an energy $E > 5 \cdot 10^{19}$ eV interact with the CMB photons, since in the proton restframe these photons appear as γ-rays, with which they interact via the so-called Δ resonance:

$$\gamma + p \rightarrow \Delta^+ \rightarrow p + \pi^0.$$

In the Δ resonance, a baryon[11] with mass $m_\Delta = 1.232\,\text{MeV}/c^2$ decays into a proton and a π^0 (or a neutron and a π^+), the resulting proton having about 20 % less energy. This process continues until the proton energy falls below the GKZ threshold. Hence, the CR spectrum should exhibit a so-called GKZ cutoff at the highest energies, which appears to have been confirmed. As a result of this (strong) interaction, the highest-energy particles cannot reach us from distances beyond about 50 Mpc, the so-called GKZ horizon.

---

[11]It consists of $\bar{u}\bar{u}d$ quarks and has spin $+3/2$.

References

Beck, R., Wielebinski, R., 2013, in *Planets, Stars and Stellar Systems*, Vol. 5, p. 641
Bethe, H.A., 1930, Ann. Phys. **397**, 325
Bietenholz, M.F., Bartel, N., Rupen, M.P., 2003, Astroph. J. **597**, 374
Braun, R., Gull, S.F., Perley, R.A., 1987, Nature **327**, 395
Chevalier, R.A., 1974, Astroph. J. **188**, 501
Crutcher, R.M., Troland, T.H., Lazareff, B., et al., 1999, Astroph. J. **514**, 121
Draine, B.T., Roberge, W.G., Dalgarno, A., 1983, Astroph. J. **264**, 485
Fermi, E., 1949, Phys. Rev. **75**, 1169
Greisen, K., 1966, Phys. Rev. Lett. **16**, 748
Han, J.L., 2007, IAU Symp. 242, p. 55
Lai, S.-P., Girart, J.M., Crutcher, R.M., 2003, Astroph. J. **598**, 392
Longair, M.S., 2011, High-Energy Astrophysics, Cambr. Univ. Press
O'Drury, L., 1983, Rep. Progr. Phys. **46**, 973
Oppermann, N., Junklewitz, H., Robbers, G., et al., 2012, Astron. Astroph. **542**, 930
Roberts, D.A., Crutcher, R.M., Troland, T.H., 1995, Astroph. J. **442**, 208
Rudnick, L., Brown, S., 2009, Astron. J. **137**, 145
Sedov, L.I., 1959, *Similarity and Dimensional Methods in Mechanics*, Acedemic Press, New York
Simard-Normandin, M., Kronberg, P.P., 1980, Astroph. J. **242**, 74
Spitzer, L., 1956, *The Physics of Fully Ionized Gases*, 2nd edition, Dover Pubn. Inc., 2006
Stahler, S.W., Palla, F., 2004, *The Formation of Stars*, Wiley, Berlin
Taylor, A.R., Stil, J.M., Sunstrum, C., 2009, Astroph. J. **702**, 1230
Uyaniker, B., Fürst, E., Reich, W., et al., 1999, Astron. Astroph. Suppl. **138**, 31
Westerhout, G., Brouw, W.N., Muller, C.A., Tinbergen, J., 1962, Astron. J. 67, 590
Wielebinski, R., Shakeshaft, J.R., Pauliny-Toth, I.I.K., 1962, The Observatory, **82**, 158
Zatsepin, G.T., Kuz'min, V.A., 1966, J. Exp. Theor. Phys. Lett., **4**, 78

Chapter 5
Magnetic Fields in Galaxies

5.1 Galaxies

The hadronic constituent of galaxies consists of gas and stars, with only a minor contribution of dust. These components rotate in a gravitational potential made up by stars and dark matter. In spiral galaxies, the gas has a neutral and an ionised phase. The neutral phase is made up by (in order of decreasing density) the molecular medium, the cold neutral medium (CNM), and the warm neutral medium (WNM). The ionised phase consists of the warm (WIM) and the hot (HIM) ionised medium (Table 5.1).

In elliptical galaxies, the bulk of the gas is ionised (HIM). An exception to this is neutral gas that has been captured from the surroundings or from neighbouring galaxies. The ionised gas in galaxies immediately implies a high conductivity, which in the absence of turbulence would result in a long-lived sustainment of magnetic fields.

Even though magnetic fields cannot have any dynamical significance on large scales,[1] they may play their role on smaller, local scales. For instance, as we have seen in Sect. 4.2.1, they must have a strong influence on the star-formation process. Another area in which magnetic fields must play a cardinal role is cosmic-ray propagation and containment within a galaxy (Sect. 5.6). Cosmic rays are produced in supernovae and are accelerated by their shock fronts. Since they consist of charged particles (p, e^-, ions), they are tightly coupled to the magnetic field, which

---

[1] There have been discussions in the literature as to whether magnetic fields could be responsible for flat rotation curves, but this is flatly impossible in view of the energy density of magnetic fields in the ISM!

© Springer International Publishing Switzerland 2015
U. Klein, A. Fletcher, *Galactic and Intergalactic Magnetic Fields*,
Springer Praxis Books, DOI 10.1007/978-3-319-08942-3_5

Table 5.1 Gas phases in galaxies, with number densities, temperatures, volume and mass filling factors, and scale heights of the disk

Gas phase	n (cm^{-3})	T (K)	f_V (%)	f_M (%)	h (pc)
Molec.	10^2–10^5	10–50	<1	~ 20	~ 70
CNM	40–80	50–200	2–4	~ 40	~ 140
WNM	0.1–0.6	5500–8500	~ 30	~ 30	~ 400
WIM	~ 0.2	~ 8000	~ 20	~ 10	~ 900
HIM	10^{-3}–10^{-2}	10^5–10^7	~ 50	~ 1	≥ 1000

in turn governs their propagation. If the CR pressure is sufficiently high, exceeding that of other pressure terms (also magnetic), they may escape from a galaxy in a galactic wind (Sect. 5.6).

In what follows, we shall discuss the properties of magnetic fields separately for spiral, dwarf irregular, and elliptical galaxies, as these have distinctly different properties concerning their kinematics or dynamics and their gas content and distribution. Spiral galaxies possess stellar and gaseous disks in which density waves can be excited, depending on the depth of their gravitational potentials. Dwarf irregulars have shallow gravitational potentials, their kinematics is therefore possibly influenced by local disturbances caused by star formation and subsequent supernovae. Large elliptical galaxies have their baryonic mass dominated by the stellar constituent, the hot gas forming a corona. They do not exhibit any systematic rotation as seen in disk galaxies. Radio-quiet ellipticals are in the focus of this chapter, i.e. those ellipticals that lack a central AGN. The radio-loud ones will be treated separately in Chap. 7, since the physics of magnetic-field creation in these is fundamentally different from that in radio-quiet ones.

5.2 Spiral Galaxies

Over the past 30 years radio continuum observations have been made of many nearby spiral galaxies, including barred galaxies, at many wavelengths. Figure 5.1 shows examples of a face-on and an edge-on galaxy; synchrotron emission and its linearly polarised component are detected across the whole disc of M 51 and over a significant distance above and below the disc of NGC 4631. The focus of these observations has always been to study, in as much depth as possible, the magnetic properties of the specific galaxy observed. Thus, we now know a lot of details about the magnetic fields of many nearby disc galaxies. There is not the space here to try to summarise all of this galaxy specific information, so we will try to provide an overview of the common magnetic properties of nearby disc galaxies. In particular we shall concentrate on those properties which are relevant to the dynamo theory discussed in Chap. 6. This will, nevertheless, result in an extensive list of references,

Fig. 5.1 *Left:* the face-on galaxy M 51 at λ6.2 cm (Fletcher et al., 2011). *Right:* the edge-on galaxy NGC 4631 at λ3.6 cm (Mora and Krause, 2013). Both images show contours of the radio continuum emission and lines showing the magnetic-field orientation, overlaid on an optical image of the galaxy (M 51: NASA/ESA/Hubble Heritage, STScI. NGC 4631: Digital Sky Survey)

which can serve as the starting point for anyone interested in tracking down details about the best studied galaxies.

5.2.1 Distribution of the Radio Emission and Magnetic Field Strengths

Spiral galaxies possess radio-emitting disks with an exponential radial decline of the brightness

$$I(r) = I(0) \exp{(-r/L)}.$$

The scale-length L is long, meaning that the radial decline in the emission is slow, with L typically about one half of the radius of the stellar disc: in other words the radio brightness at the edge of the optical disc is about 40 % of the brightness in the galaxy centre (ignoring locally strong emission in the centre due to e.g. an active nucleus). The radio emission is dominated by non-thermal (synchrotron) radiation at frequencies up to roughly 10 GHz, with thermal (free-free or bremsstrahlung) radiation dominating above this frequency. Low-mass star-forming galaxies exhibit a significantly lower fraction of synchrotron emission, most likely owing to their lower cosmic-ray containment (see Sect. 5.3).

Edge-on disc galaxies exhibit a non-thermal disk consisting of two exponential components,

$$I(z) = I_1(0) \exp\left(-\frac{z}{H_1}\right) + I_2(0) \exp\left(-\frac{z}{H_2}\right).$$

Typically the thin disc has a scale-height of $H_1 \approx 300\,\mathrm{pc}$ and thick disc has $H_2 \approx 1.8\,\mathrm{kpc}$ (Krause, 2012). Note that in the case of energy equipartition between cosmic rays and magnetic fields $I \propto B_{\mathrm{tot}}^{\delta}$ with $\delta \approx 4$, so the scale length and height of the magnetic energy density will be roughly twice those of the synchrotron emission.

In spiral galaxies the radio synchrotron spectrum (Eq. 2.27) exhibits a radial steepening from $\alpha_{\mathrm{nth}} \approx 0.6$ in galaxy centres to $\alpha_{\mathrm{nth}} \gtrsim 1.0$ at the peripheries. A similar steepening has been observed between the arm and inter-arm regions and between the disc mid-plane and the halo in a few galaxies. This spectral steepening is usually attributed to the cosmic-ray electrons losing energy, primarily via synchrotron radiation and inverse-Compton scattering off the cosmic microwave background photons, as they propagate away from their origins.

Magnetic-field strengths in spiral galaxies cannot be directly measured and are usually inferred from the synchrotron intensity (Eq. 3.13), the degree of linear polarization (Eq. 3.20) and Faraday rotation (Eq. 3.15). The synchrotron intensity allows an estimate of the total magnetic field strength B_{tot} to be made. Surveys by Hummel (1986), Fitt and Alexander (1993) and Niklas (1995) give an average total magnetic field strength of $B_{\mathrm{tot}} \simeq 10\,\mu\mathrm{G}$ for spiral galaxies, with a range of $3\,\mu\mathrm{G} < B_{\mathrm{tot}} < 20\,\mu\mathrm{G}$, using the equipartition assumption. Note that these values are derived from the mean B_{tot} for each galaxy and so there can be local regions with higher field strengths.

Once B_{tot} has been derived using the equipartition assumption, the degree of linear polarization of the synchrotron intensity, p, can be used to estimate the strength of the ordered component of the field B, where the ordering is on the scale of the resolution of the observations. If the magnetic field is perfectly ordered within the beam, i.e. $B_{\mathrm{tot}} = B$, then the degree of polarisation of the synchrotron radiation will be the theoretical maximum $p/I = p_0$ where $p_0 \approx 70\%$ is given by Eq. (2.29). If the magnetic field is perfectly disordered within the beam, i.e. $B_{\mathrm{tot}} = b$, then $p/I = 0$. More generally,

$$p = p_0 \frac{B_{\perp}^2}{B_{\mathrm{tot}\perp}^2},$$

where $B_{\mathrm{tot}\perp}^2 = B_{\perp}^2 + 2b^2/3$ if the random magnetic field is assumed to be isotropic, so $b_x^2 = b_y^2 = b_z^2 = b^2/3$. It is assumed that the reduction in fractional polarization from the maximum p_0 is solely due to emission within the beam cylinder originating at different orientations and not to the effects of Faraday rotation. In practice this means that short-wavelength ($\lambda \lesssim 6\,\mathrm{cm}$) observations are required. In both cases knowledge of the inclination of the galaxy plane to the line of sight is necessary to derive B_{tot} and B from $B_{\mathrm{tot}\perp}$ and $B_\perp$. From previously published observations of 22 galaxies Fletcher (2010) obtained an average ordered magnetic field strength of $B \approx 5\,\mu\mathrm{G}$. Since $B_{\mathrm{tot}}^2 = B^2 + b^2$ this leaves a random field of $b \approx 9\,\mu\mathrm{G}$ and $b/B \approx 2$.

Faraday rotation allows an estimate of $B_\parallel$ to be made, given some assumptions about the average electron density in and the length of the line-of-sight through the rotating region (Sect. 3.3). It is usually assumed that B lies parallel to the galactic plane (i.e. that the vertical component B_z is weak) and so B can be derived from $B_\parallel$ if the inclination of the galaxy is known. In galaxies where estimates of B have been made using both p and RM they sometimes agree, such as in M 31 (Fletcher et al., 2004), and sometimes disagree, for example in M 51 (Fletcher et al., 2011). Incompatibility between the estimates of B are due to incorrect assumptions underlying at least one of the methods followed. Three possibilities are that: (i) there is a correlation (or anti-correlation) between n_e and B, meaning that the integral describing Faraday rotation cannot simply be replaced by the product of average quantities and the line of sight path length; (ii) the assumption of equipartition of energy between cosmic rays and magnetic fields is incorrect; (iii) some fraction of the polarised emission is produced by a component of the magnetic field that is ordered, so that synchrotron emission is polarised, but does not have a consistent direction and so produces no Faraday rotation. The latter case is thought to occur in M 51 and several other galaxies, resulting in the identification of two different types of random magnetic field.

5.2.1.1 Isotropic and Anisotropic Random Magnetic Fields

For observed synchrotron radiation to be linearly polarised requires only that the plane-of-sky component of the magnetic field, $B_{\mathrm{tot}\perp}$, in the emission region, is ordered on the scale of the telescope beam. The ordering does not have to be total, what is required is enough order to prevent all of the emitted linear polarisation from within the beam mutually cancelling (see Fig. 5.2).

The order in the magnetic field can originate in at least two different ways: the magnetic field has a mean component $B_\perp$ and/or the random component of the magnetic field, b, is partially ordered. At first sight it seems strange that a random field can also be ordered! So let us briefly look at the correlation properties of a random field.

The normalised autocorrelation of a random variable, $b(x)$, is defined as

$$C(l) = \frac{1}{\langle b^2 \rangle} \langle b(l + x)b(x) \rangle,$$

where l is a separation in space and $C(l)$ is calculated for all $l > 0$. Normalisation by $\langle b^2 \rangle$ means that: $C(l) = 1$ when b is perfectly correlated over some distance l so

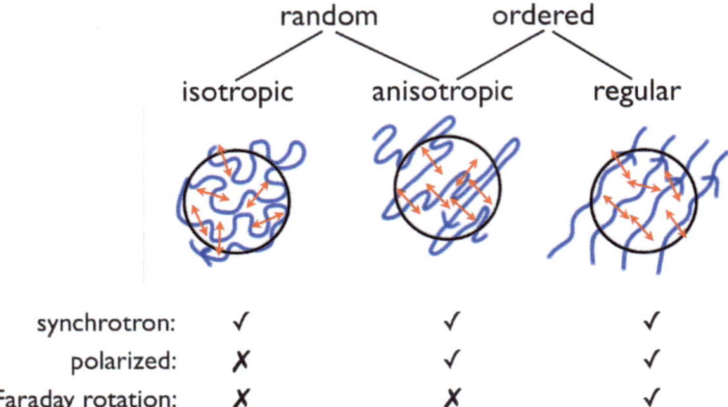

Fig. 5.2 Cartoon showing magnetic field classification. Magnetic field lines (*blue*) produce polarised synchrotron radiation with the plane of polarisation (*red arrows*) orthogonal to the field. The polarised emission is averaged over the beam area (*black*). To produce Faraday rotation the regular magnetic field (*right*) must be inclined to the line-of-sight

$b(l + x) = b(x)$ and $\langle b(l + x)b(x) \rangle = \langle b^2 \rangle$; $C(l) = -1$ for perfect anti-correlation, with $b(l + x) = -b(x)$, $\langle b(l + x)b(x) \rangle = -\langle b^2 \rangle$; and $C(l) = 0$ when there is no correlation and thus $\langle b(l + x)b(x) \rangle = 0$. A correlation length is defined,

$$l_0 = \int_0^\infty C(l)\mathrm{d}l,$$

which is a measure of the distance over which b has similar values. This can be generalised to a random variable defined in three dimensional space, such as a component of the magnetic field vector $\vec{b}(x, y, z)$ say $b_x(x, y, z)$. Now the autocorrelation is a tensor $C_{ij}(l)$ where $i, j = x, y, z$ and we have correlation lengths l_{0xx}, l_{0xy}, etc. An isotropic random field has the same properties irrespective of direction and so all of the correlation lengths will be the same. If the correlation length in one direction is larger than the others, the field maintains its coherence in that direction for a longer distance: such a field is not isotropic, as there is a preferred direction, and is called anisotropic. The extended correlation length produces the order in the magnetic field that results in a linearly polarised component of the synchrotron emission.

An alternative way in which to define anisotropy is to compare the variances of the random magnetic field components. An isotropic field, in Cartesian coordinates, has $\sigma_x^2 = \sigma_y^2 = \sigma_z^2$, where the variance in b_x is $\sigma_x^2 = \langle b_x^2 \rangle - \langle b_x \rangle^2$, with similar equations for the variances in b_y and b_z. Suppose our plane-of-sky coordinates are x and y, and the random field is anisotropic, so $\sigma_x^2 \neq \sigma_y^2$. The observed degree of polarisation is (Sokoloff et al., 1998)

Table 5.2 Galaxies with anisotropic random magnetic fields

Galaxy	Analysis	Reference
Milky Way	$\sigma_{\mathrm{RM}} \propto \mathrm{RM}$	Brown and Taylor (2001)
NGC 1097 & NGC 1365	I & p weakly amplified by bar shock	Beck et al. (2005)
Milky Way	Modelling I, p & RM in Galactic plane	Jaffe et al. (2010)
M 51	Strong p but weak RM	Fletcher et al. (2011)
Milky Way	Modelling I, p & RM across whole sky	Jansson and Farrar (2012)
M 51	Dispersion of polarisation angles	Houde et al. (2013)

Notes: RM and σ_{RM} are the Faraday rotation measure and its standard deviation, respectively. I and p are the synchrotron intensity and its fractional polarization, respectively

$$\frac{p}{p_0} = \frac{|\sigma_x^2 - \sigma_y^2|}{\sigma_x^2 + \sigma_y^2}, \tag{5.1}$$

which gives $p = 0.13 p_0 \approx 0.1$ when $\sigma_x/\sigma_y = 1.14$, i.e. for a weakly anisotropic random magnetic field the synchrotron emission will be about 10 % polarised. Significant p can thus be produced in the absence of any regular magnetic field.

Figure 5.2 sketches three different components of the magnetic field – regular and anisotropic and isotropic random fields – and summarises how these components are related to observable quantities.

Table 5.2 lists the galaxies where the presence of anisotropic random magnetic fields has been inferred from observations. Although only four different galaxies are listed, different observational signatures have been interpreted to reveal anisotropic fields where a galaxy appears more than once in the table. This is reassuring, in that the identification of anisotropic random fields does not rely upon a single approach to data analysis.

The physical origin of anisotropy in b is not yet clear and may even be different in different galaxies. There are at least three plausible mechanisms. Firstly, compression of an initially isotropic random magnetic field by a shock. A shock front passing through a magnetic field amplifies the components of the magnetic field which are parallel to the shock, but does not affect perpendicular components. This selective amplification results in an anisotropic magnetic field lying downstream of the shock, even though the field upstream is isotropic. Supernova remnants (SNR) expand with a supersonic velocity into the ISM and density waves can cause large-scale shocks in the ISM as it crosses spiral arms, away from the co-rotation radius: both are thus potential sources of anisotropic random magnetic fields. Since SNR shocks should be randomly distributed, we expect that they would result in a random pattern of synchrotron polarisation orientations, whereas shocks along spiral arms would tend to line up the magnetic field with the arm over large distances, rather similar to the patterns of polarisation that are observed. This effect has been clearly demonstrated in M 51 (Patrikeev et al., 2006) and may occur in other galaxies. Secondly, shearing of an isotropic random magnetic field by the galactic differential rotation will stretch the field components in the direction of the rotation. This will produce a random

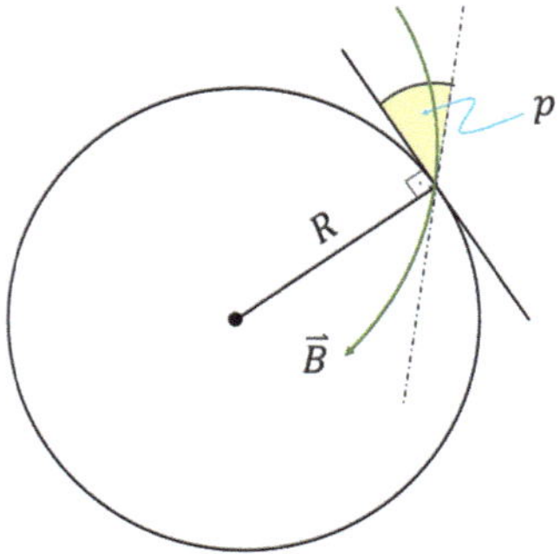

Fig. 5.3 Pitch angle of a spiral: this can be a spiral arm or a spiral magnetic field line

magnetic field that has a longer correlation length in one direction and so again will result in linearly polarised synchrotron emission even in the absence of a large-scale, regular magnetic field. Finally, MHD turbulence itself is inherently anisotropic, with a longer correlation length parallel to any background mean field.

5.2.2 Magnetic Field Pitch Angles

One of the most useful parameters of a galactic magnetic field is its pitch angle,

$$p_{\mathrm{B}} = \arctan\left(\frac{B_r}{B_\phi}\right),\tag{5.2}$$

which measures the orientation of the magnetic field with respect to the local line of fixed galacto-centric radius (the local galacto-centric circumference), shown in Fig. 5.3. Pitch angles are easier to measure than magnetic field strengths since the plane-of-sky magnetic field orientation is directly obtained from the observed Stokes parameters Q and U using Eq. (3.8). If the inclination of the galaxy disc to the line-of-sight is known, p_{B} in the galactic plane can be recovered using simple geometry.

In spiral galaxies the magnetic field lines, traced by short-wavelength or Faraday rotation corrected polarisation angles,[2] trace similar spiral lines as the optical or gaseous spiral arms (Fig. 5.1). Table 5.3 compares the measured magnetic and spiral arm pitch angles for six galaxies and shows that there is often a close agreement between the two measures.

The spiral arm pitch angles only exist at the location of the arms, whereas p_{B} can be determined wherever sufficient signal in Q and U can be measured. The reason for the similarity between the two pitch angles is not fully understood.

---

[2]The observed polarisation angles, given by Eq. (3.8), are oriented parallel to the electric field. These are rotated by 90° to obtain the magnetic field orientation, sometimes called the "B-vectors".

Table 5.3 Pitch angles of the magnetic field and spiral arms in some spiral galaxies

Galaxy	Regular magnetic field		Spiral arms	References	
	Inner	Outer		Magnetic field	Spiral arm
IC 342	$-20 \pm 2°$	$-16 \pm 2°$	$-19 \pm 5°$	[1]	[7]
M 31	$-17 \pm 4°$	$-8 \pm 3°$	$-7°$	[2]	[8]
M 33	$-48 \pm 12°$	$-42 \pm 5°$	$-26 \pm 5°$ [a]	[3]	[9]
M 51	$-20 \pm 1°$	$-18 \pm 1°$	$-15 \pm 8° \rightarrow -10 \pm 8°$	[4]	[10]
M 81	$-14 \pm 7°$	$-22 \pm 5°$	$-11° \rightarrow -17°$ [b]	[5]	[11]
NGC 6946	$-38 \pm 12°$ [c]		$-36 \pm 7°$ [c]	[6]	[6]

Negative pitch angles denote trailing spirals

[a] Average value for 10 spiral arm segments

[b] Change of spiral arm pitch angle from inner to outer galaxy

[c] Average values for 5 spiral arms that have different radial extents

References: [1] Gräve and Beck (1988); [2] Fletcher et al. (2004); [3] Tabatabaei et al. (2008);
[4] Fletcher et al. (2011); [5] Krause et al. (1989); [6] Frick et al. (2000); [7] Kennicutt (1981);
[8] Nieten et al. (2006); [9] Sandage and Humphreys (1980) (note that this paper & the erratum
define the pitch angle with respect to the radial line rather than the more usual circumference); [10]
Berkhuijsen et al. (1997); [11] Oort (1974)

A large-scale spiral shock, due to a density wave origin for the spiral arms, would
align the magnetic field and arm at the position of the arms but in the inter-arm
region p_B could vary. This pattern has been shown in the case of M 51 by Patrikeev
et al. (2006). However, it is not clear that all of the galaxies in Table 5.3 host density
waves, neither is it obvious why the *average* p_B is so similar, in most cases, to the
spiral arm pitch angle.

Where available, two different p_B values are given for the galaxies in Table 5.3.
This is because estimates of p_B from dynamo theory, described in Sect. 6.4.2.3,
predict that the pitch angle of the field should decrease with increasing radius if
the gaseous disc is flared (as is the case for the Milky Way and likely other spiral
galaxies.)

Pitch angles are important diagnostics of the magnetic field as they are a natural
output of the dynamo models which are discussed in Chap. 6, thus allowing the
theory to be directly compared to observations.

5.2.3 Magnetic Field Symmetry

The symmetry properties of regular magnetic fields, with respect to both rotation
around the disc axis and reflection about the disc mid-plane, are also closely connected with the mean-field galactic dynamo theory discussed in Chap. 6. Regular-
field symmetry is much more difficult to obtain from observations than the magnetic
pitch angles, particularly the vertical symmetry of the field components, but some
progress has been made, which we summarise here.

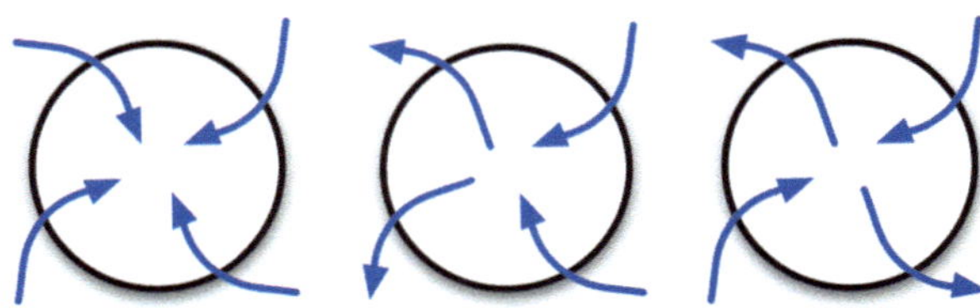

Fig. 5.4 Cartoon showing a face-on view of: axisymmetric $m = 0$ (*left*), bisymmetric $m = 1$ (*centre*) and $m = 2$ (*right*) regular magnetic fields

Historically, two fundamental magnetic-field configurations describing the azimuthal pattern of the regular field were identified. In each configuration the field is assumed to have the same pitch angle around a circle of fixed radius from the galactic centre. In one case the direction of the regular field is the same around the circle (this direction can be either inward or outward pointing, there is no preference); since this field is perfectly symmetric for any rotation about the galactic centre it is called axisymmetric. In the other case the field direction changes sign, pointing inwards around half of the circle and outwards around the other half; this type of field is called bisymmetric. Figure 5.4 illustrates these two regular field patterns.

The motivation for the axisymmetric/bisymmetric field classification is that it provides a simple way to distinguish between two proposed origins for the observed regular fields: the mean-field dynamo and a wound-up primordial field. Distinguishing between these theories is discussed in detail in Chap. 6, here we concentrate on how to recognise these patterns in observational data.

Let us assume that the galactic magnetic field and the distribution of thermal electrons give rise to a (constant) rotation measure RM_0 if we were looking directly along the field. Observing the galaxy from afar, when it is inclined by an angle i to our line-of-sight (where $i = 0$ corresponds to the face-on view) the rotation measure as a function of azimuthal angle ϕ will vary as

$$\mathrm{RM}(\phi) = \mathrm{RM}_0 \tan i \, \cos(\phi - p_\mathrm{B}),$$

in the case of an axisymmetric field. Over a complete ring around the galactic centre, $0 < \phi < 2\pi$, this equation is singly periodic in ϕ. In the case of a bisymmetric structure,

$$\mathrm{RM}(\phi) = \frac{1}{2}\mathrm{RM}_0 \tan i \, [\cos(2\,\phi - p_\mathrm{B} - \mu) + \cos(p_\mathrm{B} - \mu)],$$

where μ is the position angle where the magnetic field reverses its direction, i.e. where $\vec{B} = 0$. Thus the observed azimuthal variation in RM, in principle, allows one to distinguish axisymmetric and bisymmetric regular magnetic field patterns.

This approach can be generalised by expanding the field in a Fourier series of the form:

$$\vec{B}(\vec{r}, t) = \sum_m B_m(r, z, t) \exp(i\, m\, \phi) \tag{5.3}$$

Table 5.4 Azimuthal modes of the regular magnetic field in galactic discs

| Galaxy | Relative mode amplitude | | | References |
	$m = 0$	$m = 1$	$m = 2$	
IC 342	1	–	–	Krause et al. (1989)
LMC	1	–	–	Gaensler et al. (2005)
M 31	1	0	0	Fletcher et al. (2004)
M 33	1	1	0.5	Tabatabaei et al. (2008)
M 51	1	0	0.5	Fletcher et al. (2011)
M 81	–	1	–	Krause et al. (1989)
M 81	0.5	1	–	Sokoloff et al. (1992)
NGC 253	1	–	–	Heesen et al. (2009)
NGC 1097	1	1	1	Beck et al. (2005)
NGC 1365	1	1	1	Beck et al. (2005)
NGC 4254	1	0.5	–	Chyży (2008)
NGC 4414	1	0.5	0.5	Soida et al. (2002)
NGC 6946	1	–	–	Ehle and Beck (1993)

Note: The relative mode amplitudes are approximate, see the relevant papers for the actual amplitudes. A dash indicates that combinations of modes including this entry were not sought whereas a zero indicates that the mode was searched for but was not found. Only galaxies where a model has been fitted to the observations are included; in the case of M81 two different models gave different results so both are included (Based on Fletcher (2010))

where $m \geq 0$ is an integer. This expansion defines a hierarchy of azimuthal regular magnetic field modes, with each mode identified by m. In this system $m = 0$ is the axisymmetric mode, $m = 1$ the bisymmetric, $m = 2$ a mode with two peaks and troughs, and so on. Rather than trying to identify patterns in RM, which is already two-steps removed from the observational data Q and U, a method developed by Sokoloff et al. (1992) and Berkhuijsen et al. (1997) for fitting a model field expansion of the form given by Eq. (5.3) directly to polarisation angles is often used.

Table 5.4 shows that the axisymmetric $m = 0$ mode has been identified in the discs of all of the galaxies where modelling of the regular magnetic field, often in terms of an expansion like Eq. (5.3), has been attempted. This is some of the strongest evidence that a mean-field dynamo is the origin of the regular fields; the connection between Table 5.4 and the theory is discussed in Sect. 6.4.3.

In contrast to disc fields, the properties of magnetic fields in galaxy haloes are more poorly known. In galaxies viewed face-on their study is especially difficult and requires very careful analysis of Faraday rotation. One galaxy where this approach has been used is M 51. Fletcher et al. (2011) identified a regular magnetic field configuration in the halo that has a dominant $m = 1$ azimuthal mode, in contrast to the disc field which is dominated by $m = 0$.

When viewed edge-on, the linear polarisation observed from disc galaxies is accumulated along the entire lines of sight above and below the disk, which

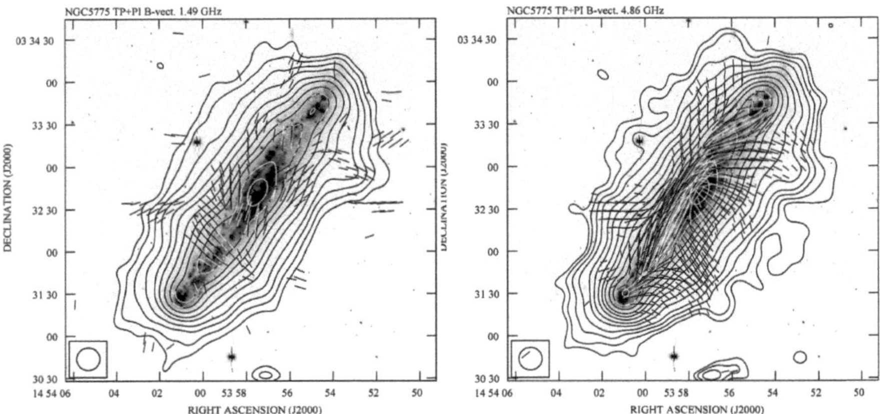

Fig. 5.5 Magnetic field in the edge-on galaxy NGC 5775 (Soida et al., 2011). Contours show the total (*left*) and polarised (*right*) radio continuum emission at $\lambda = 6.2$ cm, the bars denote the magnetic-field orientation

can hamper its interpretation. Mildly inclined galaxies such as M 31 provide the advantages of face- and edge-on views in one observation.

Edge-on galaxies typically show the ordered magnetic field to be plane-parallel to the disc within the disc, while it has a vertical component at larger heights. Typical examples are NGC 4631 (Fig. 5.1) and NGC 5775 (Fig. 5.5), with radio continuum spurs associated with star-forming regions in the underlying disc. The magnetic field in the spurs is probably dragged outward by a galactic wind. Galaxies with massive discs and normal star formation rates exhibit X-shaped magnetic-field morphologies in their haloes (NGC 253, NGC 891, NGC 3667, NGC 4565, NGC 5775). Whether this is connected with field topologies resulting from a galactic dynamo or just mimics the flaring of the outer gaseous disks is, at present, unknown.

It is often not clear if the linear polarisation detected in edge-on galaxies is produced by an ordered or a regular magnetic field. It is difficult to interpret Faraday rotation measurements, due to the extended lines-of-sight, and so the only method for determining the magnetic field direction is difficult to apply. In the case of NGC 891 (Krause, 2009) and NGC 5775 (Fig. 5.5) (Soida et al., 2011) the rotation measure of the diffuse polarised emission is the same above and below the disc midplane, a strong indication that the radial and azimuthal components of the regular magnetic field, B_r and B_ϕ, have the same sign for $z > 0$ and $z < 0$. Mao et al. (2012) obtained the same symmetry property for the Large Magellanic Cloud, using diffuse polarisation and background rotation measures. These few observational determinations of the vertical field symmetry are favourable for the mean-field galactic dynamo theory (see Sect. 6.4.2.2).

An exceptional case is the starburst galaxy M 82. Its synchrotron brightness implies $B_t \approx 50\,\mu$G in its central 700-pc area! This is the result of its intense star formation and supernova rate. Not unexpectedly, its magnetic field exhibits a vertical or radial orientation. This brings us to a particular class of disk galaxies, which host forming or active supermassive black holes in their centres (Fig. 5.6). Superimposed onto their 'normal' radio disks, they exhibit rather well-focused winds from their centres in the first case, or even a central source, twin jets, and radio lobes in the latter – just like radio galaxies (Sect. 7.4). The most prominent examples are complied below (see Sect. 7.1 for the AGN taxonomy).

Galaxy	AGN type	Radio features
0313-192	Sy 2	FRI
0421+040	Sy 2	FRII (?)
NGC 2992	Sy 2	'Figure 8'
NGC 3079	LINER	'Figure 8'
NGC 3367	LINER	FRII
NGC 3516	Sy 1	one-sided, curved, blobs
NGC 4258	Sy 2	'anomalous radio arms'; ejection in disk

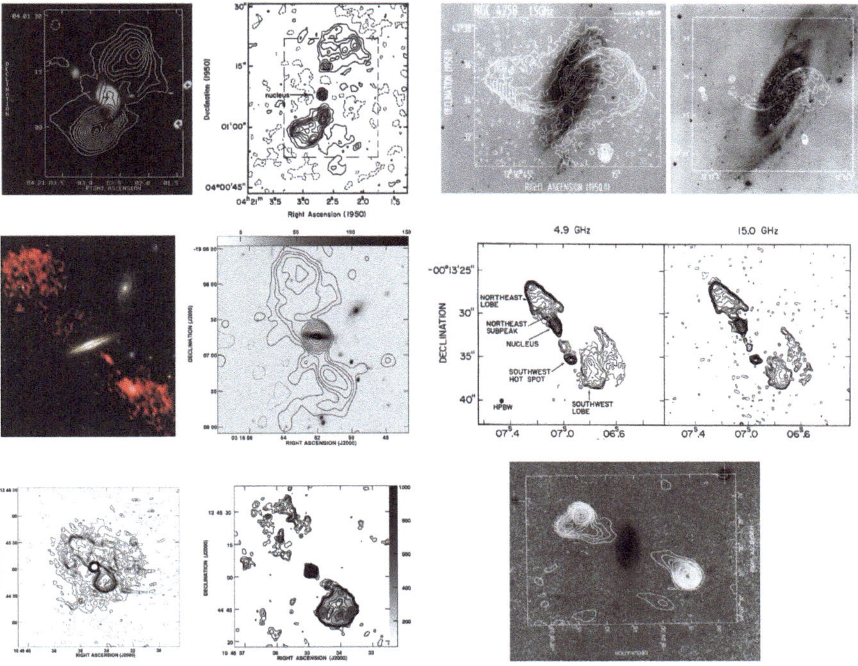

Fig. 5.6 Potpourri of spiral galaxies with central AGN and radio-galaxy features. Shown are: 0421+040 (*top left*), NGC 4258 (*top right*), 0313-192 (*middle left*), NGC 1068 (*middle right*), NGC 3367 (*bottom left*), 0400-181 (*bottom right*)

As we shall see in Sect. 7.4, there are two classes of radio galaxies, namely so-called FR I and FR II objects (Sects. 7.4.2 and 7.4.1), which are usually hosted by elliptical galaxies, i.e. galaxies lacking any cold neutral gas, but having only hot, X-ray-emitting gas. It is most surprising to find these phenomena also in some spiral galaxies. The magnetic-field strengths in the anomalous radio features of spiral galaxies are ≥ 10 times higher than those responsible for the normal disk emission. Seyfert-type galaxies are likely to possess a central black hole with an accretion disk, from which twin jets are launched, probably at a large angle with respect to the disk. An exception to this is probably NGC 4258.

So-called 'Figure-8' structures are the result of focussed winds from their active central regions. In these galaxies, a wind with a speed of $v \approx 5000\,\mathrm{km\,s^{-1}}$ is focussed by the disk while ploughing through the ISM 'above' and 'below' the central active region. The magnetic-field strength in these somewhat collimated structures is between several tens and a few hundred μG. There are many more spirals with Seyfert or LINER activity (see Sect. 7.6). In most Seyferts, the central AGN is not powerful enough to produce any FRI/II morphology, as the jets are quenched by the dense central ISM, so that such structures are inhibited, and the synchrotron luminosities not correspondingly enhanced.

5.3 Dwarf Irregular Galaxies

Low-mass (or dwarf) galaxies are mostly weaker radio emitters, even though some of them, being in a phase of intense star formation, exhibit high radio surface brightnesses. If they do, they emit radio (synchrotron and free-free) emission during their star-formation periods, and for some time thereafter. Measurements of the radio continuum radiation of dwarf galaxies over a large frequency range have shown that ongoing star formation in them is accompanied by enhanced radio continuum emission, but with a significantly lower fraction of synchrotron radiation as compared to massive spiral galaxies. This is most likely connected to their small mass. With their shallow gravitational potentials, their escape speeds are much lower ($\sim 1/4 \ldots 1/3$) than in massive spirals, hence they are prone to mass loss while experiencing strong bursts of star formation. Such galactic winds not only enrich the intergalactic space (ICM, IGM) with heavy elements ('metals'), but at the same time they transport relativistic particles and magnetic fields out of these galaxies into the surroundings. The majority of dwarf galaxies is seen in quiescent phases though, hence they are weak radio emitters. Our knowledge of the magnetic-field properties in low-mass galaxies is accordingly poor, resting upon but a handful of them. The compilation given further below contains in brief our current state of knowledge.

The magnetic-field structures in dwarf galaxies are much less ordered than in massive spirals, which is probably due to their slow rotation. In Fig. 5.7, two prominent examples are shown, viz. NGC 1569 and NGC 4449. These are starburst dwarf galaxies producing strong radio emission (both thermal and nonthermal). Their intense synchrotron radiation facilitates the detection of the polarised component

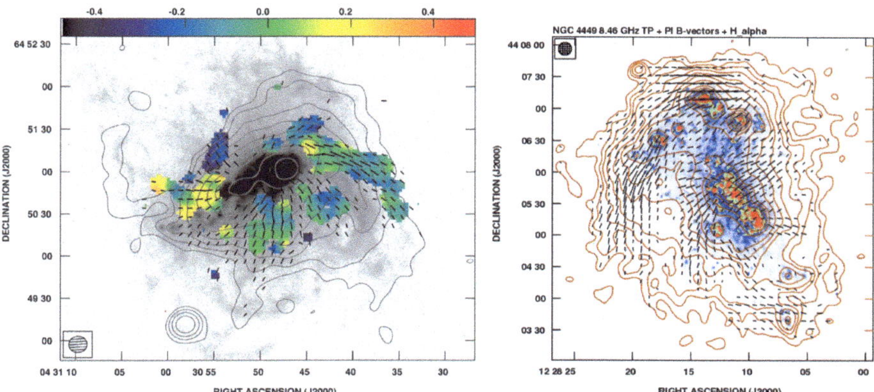

Fig. 5.7 Examples of magnetic fields in the dwarf galaxies. *Left*: radio continuum image of NGC 1569 at 8.5 GHz (contours), superimposed onto an Hα image (*gray scale*). The magnetic field is indicated by the *black bars*, and the rotation measure by the colour scale, with the wedge representing units of 1000 rad m^{-2} (Kepley et al., 2010). *Right*: Radio continuum map of NGC 4449 at 8.5 GHz (contours), superimposed onto an Hα image (colour). The magnetic field is again indicated by *bars* (Chyży et al., 2000)

and hence to study their magnetic fields. Note that most dwarf galaxies are radio-faint, rendering the detection of magnetic fields in them very difficult.

In spite of their mostly rigid rotation, however, dwarf galaxies are able to produce large-scale, coherent magnetic-field patterns. Non-standard dynamos may be at work in them. Some information is available for our nearest galactic neighbours. In the Small Magellanic Cloud (SMC), the line-of-sight component of the magnetic field appears to point uniformly away from us, while the projected field points roughly in an east-western direction. There have been speculations whether this hints at a so-called 'pan-magellanic' field connecting the LMC and the SMC. Also in case of the SMC, a CR-driven dynamo seems to be favoured by the observations. For the prototypical starburst dwarf galaxies NGC 4449 and NGC 1569, the discussion is whether a dynamo, perhaps supernova-driven, or a galactic wind is shaping the magnetic fields.

Dwarf galaxies may have been, or could still be, important in 'magnetising' the ICM/IGM, i.e. transporting relativistic particles and magnetic fields into their surroundings. Spectroscopic observations of emission lines (Hα, [NII]) indicate wind speeds reaching, or exceeding, the escape speed

$$v_{esc} = \sqrt{\frac{2\,G\,M}{R}}.$$

(5.4)

By the same token, the measured temperatures of the hot (X-ray-emitting) gas reaches, or exceeds, the virial temperature

Galaxy	B_t	B_u	Remarks
LMC	6 μG (equipart.)	2 μG (filaments)	Diffuse pol. emission RM grid (backgroud sources) of 192 accurate RMs
SMC	4 μG (equipart.)	1.7 μG	RM grid; optical pol.; 70 accurate RMs
IC 10	14 μG	2–3 μG	B_u associated with giant bubble
NGC 1569	39 μG (centre) 10–15 μG (halo)	3–9 μG (strongest in halo)	Radial field, synchrotron halo
NGC 4449	14 μG	8 μG	Radial, with spiral structure; counter-rotation; synchr. halo
NGC 6822	$\leq$5 μG	2–3 μG	B_u just indicated (polarised emission)

$$T_{vir} = 1.5 \cdot 10^5 \cdot \left(\frac{v_{esc}}{100 \, \text{km s}^{-1}} \right) \text{K.} \qquad (5.5)$$

It is therefore quite likely that magnetic fields and relativistic particles have been ejected into the ICM/IGM by low-mass galaxies, in particular soon after their formation in the young universe. They were numerous in a ΛCDM cosmology, they had low escape speeds, hence a low containment of CR (Sect. 5.6). This process must have been competing with magnetisation by AGN (Chap. 7).

5.4   Elliptical Galaxies

Elliptical galaxies, and likewise dwarf ellipticals and spheroidals, are fundamentally different from disk galaxies, also as far as their gaseous content is concerned. They lack the cold gas, but rather have an ISM of hot $\geq 10^6$ K gas. Dwarf ellipticals and spherodials may not have any gas at all, as their gravitational potentials may not be deep enough to retain it. The obvious reason for this is that their evolution is 'completed', in the sense that their stars are evolved and their cold gas has been consumed in star formation, or expelled. The hot gas needs a long time to cool,

$$\tau_{cool} = 1.2 \cdot 10^6 \cdot \left(\frac{n_p}{\text{cm}^{-3}} \right)^{-1} \cdot \left(\frac{T}{10^6} \, \text{K} \right)^{\frac{1}{2}} \text{year} \qquad (5.6)$$

so that star formation is inhibited. Without star formation, there will not be any related radio emission, due to the lack of production of relativistic particles. By the same token, there are no sources for magnetic fields to be generated and/or sustained. The radio continuum emission from elliptical galaxies is hence confined to the central regions of those ellipticals that are sufficiently massive, so that sufficient amounts of cooled gas may feed a central machine, viz. an AGN (Chap. 7). Diffuse, extended emission is therefore not seen in ellipticals.

Hence, elliptical galaxies only possess central synchrotron-emitting sources, if any. These may be AGN producing jets and lobes (such as in radio galaxies), or are just confined compact sources, probably because the surrounding ISM is too dense, and the AGN too weak, for the jets to make it out of the galaxy. Some properties may be worth mentioning here:

- A correlation exists between the isophotal shape of ellipticals and their radio luminosity. Ellipticals with boxy or irregular optical morphologies are radio-loud and tend to possess X-ray haloes, whereas galaxies with pointed isophotes[3] are radio-quiet and do not show any X-rays.
- A radio-X-ray correlation exists, but has a large scatter. This hints at an accretion flow to fuel central radio sources.
- Ellipticals possess, if any, mostly unresolved central radio sources. Some show extended (linear) features, similar to the central sources in star-forming spiral galaxies.
- Most ellipticals depart from the radio-FIR correlation established for normal disk galaxies. The excess radio emission is obviously produced by the AGN, hence a different mechanism for the particle production and acceleration has to be envisaged here.
- The extended radio lobes of elliptical galaxies hosting an AGN come along with low X-ray to optical ratios. This hints at a more tenuous ISM, which is less impeding to the formation of radio jets.

5.5 The Radio-FIR Correlation

The production of relativistic particles and magnetic fields is closely linked to star formation activity in their disk, with massive stars ionising the gas surrounding them and subsequent supernova explosions injecting and accelerating particles, mainly protons and electrons. The UV radiation of the stars is very efficiently absorbed by the dust and reradiated in the FIR regime, this naturally explaining the well-established radio-FIR correlation, discovered coincidentally in the same year by two groups (de Jong et al., 1985; Helou et al., 1985) see Fig. 5.8.

Note that this correlation was established utilising radio continuum measurements at frequencies at which the nonthermal emission dominates, so it is one between the (nonthermal) synchrotron radio emission and the (thermal) FIR. The correlation between the thermal (free-free) radio and the FIR emission is immediately expected and trivial, since both constituents reflect nothing but the

---

[3]The term 'pointed' means weak edge-on disks.

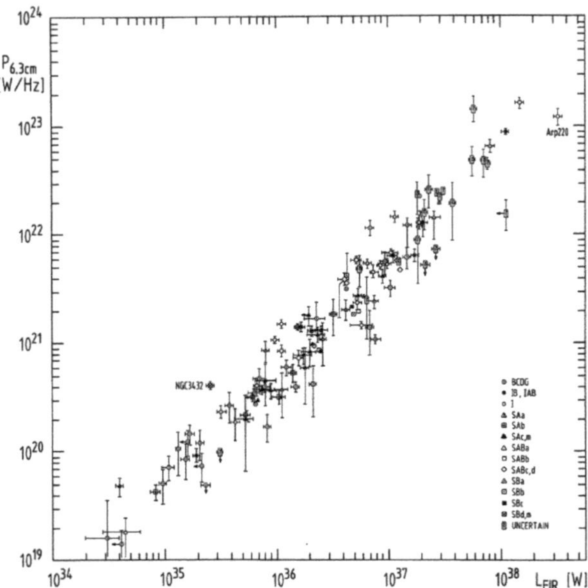

Fig. 5.8 Radio-FIR correlation of star-forming galaxies (Wunderlich et al., 1987)

luminosity of Lyman continuum photons in each galaxy: these photons ionise the gas, giving rise to the thermal radio continuum, until they are ultimately absorbed, their energy being reradiated in the FIR. Both luminosities are therefore directly proportional to the underlying UV luminosity, hence the star formation rate. The fact that the radio-FIR correlation also holds for the nonthermal radio continuum is still due to the star-formation rate, more specifically, to that integrated over the past $\sim 10^8$ year. The reason is that the relativistic particles produced by the supernove are stored within the galaxies for some time and are kept relativistic, owing to the ongoing supernova activity and Fermi acceleration processes. This is the reason for this correlation being so tight for normal star-forming galaxies as evident in Fig. 5.8.

5.6 CR Containment

Relativistic particles (e^-, p, ...) are charged, hence tightly coupled to the magnetic field, The latter in turn is coupled to the gas via the ionised (thermal) component. Hence, all of these constituents form a disk (with different scale heights) which is subject to hydrostatic equilibrium unless it is overpressured by strongly enhanced star formation and subsequent supernova activity. This process may lead to a galactic wind in which the relativistic particles and magnetic fields are transported out of a galaxy into intergalactic space. Without such convective transport, the propagation of relativistic particles is subject to diffusion, in which they experience multiple scattering from Alfvén waves, thereby moving at the Alfvén velocity

$$v_A = \frac{B}{\sqrt{4\,\pi\,\rho_{ion}}},$$ (5.7)

where ρ_{ion} is the density of the ionised gas. Supposing that the galaxy disk is pervaded by a uniform, plane-parallel magnetic field, hydrostatic equilibrium in the z-direction (perpendicular to the disk) is given by

$$\frac{dP}{dz} = -\rho \cdot \frac{d\Phi}{dz}.$$ (5.8)

The density ρ is related to the gravitational potential by the Poisson equation

$$\frac{d^2\Phi}{dz^2} = 4\,\pi\,G\,\rho.$$ (5.9)

For $|z| \leq 250$ pc, the gravitational acceleration a_z has a roughly constant slope (Fig. 5.9) so that we can write

$$\frac{d\Phi}{dz} = -z \cdot \frac{da_z}{dz}.$$ (5.10)

Hence, from (5.8) we see that

$$\frac{dP}{dz} = \rho\,z \cdot \frac{da_z}{dz}$$ (5.11)

The pressure is the sum of pressure contributions from the gas, cosmic rays and the magnetic-field. The gas pressure is dominated by macrosopic turbulence rather than the thermal one, so that

$$P_g = \frac{1}{3} \cdot \rho \left\langle v^2 \right\rangle = 1.0 \cdot 10^{-12}\ \mathrm{dyn\,cm}^{-2}$$ (5.12)

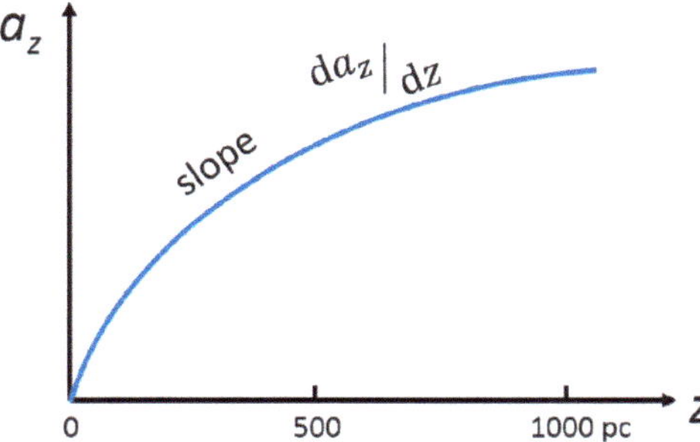

Fig. 5.9 Sketch of the gravitational acceleration as a function of height above the galactic disk

The CR pressure is one third of the energy density of the relativistic particles, $u_{CR} \approx 1.3 \cdot 10^{-12}\,\mathrm{erg\,cm}^{-3}$, i.e. $P_{CR} = 0.4 \cdot 10^{-12}\,\mathrm{dyn\,cm}^{-2}$. The magnetic pressure is

$$P_{mag} = \frac{B^2}{8\,\pi} = 1.0 \cdot 10^{-12}\,\mathrm{dyn\,cm}^{-2} \tag{5.13}$$

for a magnetic-field strength $B = 5\,\mu\mathrm{G}$. Now, assuming that the three pressures vary with height z above (and below) the galactic plane in the same way, i.e.

$$P_{mag} = \alpha \cdot P_g$$

$$P_{CR} = \beta \cdot P_g$$

the equation for hydrostatic equilibrium reads

$$(1 + \alpha + \beta) \cdot \frac{1}{3\,\rho} \cdot \frac{d}{dz}\left(\rho \langle v^2 \rangle\right) = z \cdot \frac{da_z}{dz}. \tag{5.14}$$

Assuming that $\langle v^2 \rangle$ is constant, the solution is

$$\rho(z) = \rho(0) \cdot e^{-\left(\frac{z}{h}\right)^2}, \tag{5.15}$$

where

$$h = \left[\frac{2 \cdot (1 + \alpha + \beta)\,\langle v^2 \rangle}{-3 \cdot \frac{da_z}{dz}}\right]^{\frac{1}{2}} = 100 \cdot (1 + \alpha + \beta)^{\frac{1}{2}}\,\mathrm{pc}. \tag{5.16}$$

is the scale height of the disk. One usually quotes the effective thickness, defined as the total mass per unit area, divided by $\rho(0)$:

$$2\,H = \frac{\int_0^\infty \rho(z)\,z\,dz}{\int_0^\infty \rho(z)\,dz} = \pi^{\frac{1}{2}}\,h = 180 \cdot (1 + \alpha + \beta)^{\frac{1}{2}}\,\mathrm{pc}. \tag{5.17}$$

Inserting $\alpha = 1$ and $\beta = 0.4$, we find $2\,H = 280$ pc. Without any enhanced star formation, hence supernova activity, the disks of galaxies would attain a more or less constant thickness of the above order. The CR particles propagate through the disk via diffusion. The discussion of CR propagation is complicated, and a thorough treatment would be beyond the scope of this textbook. We therefore indicate the issues of interest with a few simplified arguments. The morphology of magnetic fields in interstellar or intergalactic space is never such that they are homogeneous. Even if they are, the streaming motion of the relativistic particles imposes fluctuations onto the fields. If the size scale of the fluctuations is much smaller than the Larmor radii of the particles, then these will just move on their helical orbits in the mean field. However, if the Larmor radii are smaller than the fluctuation scales, then the relativistic particles will experience frequent pitch-angle changes, i.e. they will be permanently scattered by the irregularities of the magnetic field. This readily explains their isotropic arrival directions. Without derivation, the mean free path of this scattering process is

$$\lambda_{sc} \approx r_g \cdot \left(\frac{B_0}{B_1}\right)^2, \tag{5.18}$$

where r_g is the Larmor radius of the particles, B_0 is the mean strength of the magnetic field, and B_1 is the strength of the perturbations inherent to this field. The diffusion coefficient then is

$$D \approx v \cdot \lambda_{sc}. \tag{5.19}$$

The irregularities can be considered as transverse waves propagating along the field lines. The bulk streaming motion of the charged particles is limited by the Alfvén speed

$$v_A = \frac{B}{\sqrt{4\,\pi\,\rho_{ion}}} = 2.2 \cdot \left(\frac{B}{\mu G}\right) \cdot \left(\frac{n_e}{cm^{-3}}\right)^{-\frac{1}{2}} \ km\,s^{-1}, \tag{5.20}$$

owing to the multiple scattering off the field irregularities. This scattering also implies an efficient 'storage' of the CR particles in a galactic disk that they are confined in. Their containment is described by the diffusion loss equation:

$$\frac{dN(E)}{dt} = \frac{d}{dE}\left[b(E)\,N(E)\right] + Q(E,t) + D \cdot \vec{\nabla}^2 N(E), \tag{5.21}$$

where

$$b(E) = -\frac{dE}{dt}. \tag{5.22}$$

describes the energy losses (synchrotron, inverse-Compton, adiabatic,...). $Q(E,t)$ is a source term, since the particles may be continuously injected (per unit volume), and the last term on the right-hand-side stands for the diffusion losses, i.e. the term quantifying the number of particles entering and leaving a volume dV by diffusion (Longair, 2011).

What happens in a real disk? In a real disk, there is star formation, producing energetic (long-lasting) stellar winds, due to (massive) star formation, followed by supernova explosions, which are even more energetic. They inject both, energetic particles as well as mechanical energy, i.e. they produce turbulence. As a result, the

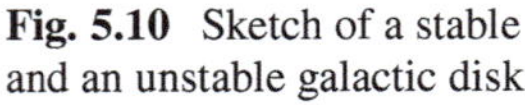

Fig. 5.10 Sketch of a stable and an unstable galactic disk

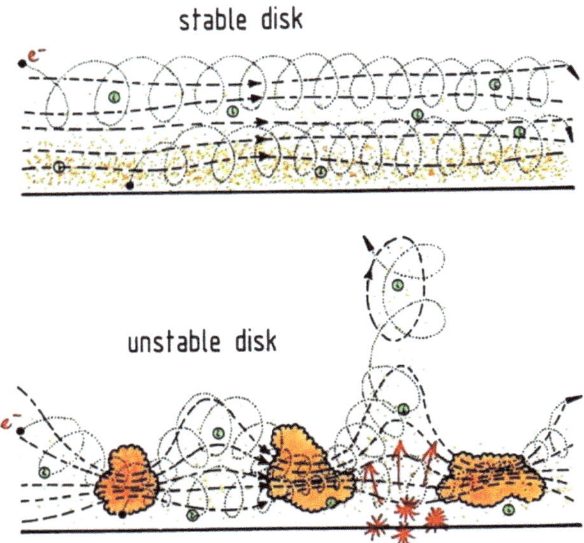

pressure increases so that the disk thickness in the vertical (z-) direction is strongly increased (Fig. 5.10). The formation of dense clouds, requiring cloud contraction and collapse, implies that under magnetic-flux conservation the magnetic-field strength is enhanced within the denser regions, while in between it is diluted.

This is the famous *Parker instability* (Parker, 1966) producing so-called 'Parker Loops' (Fig. 5.10). Eventually, when there is star-formation activity between the clouds (with some time delay), the particles and fields may be pushed out of the disk, whereby magnetic-field lines may reconnect to form buyoant loops, transporting magnetic fields and particles away from the disk into the galaxy halo and, if sufficient energy is injected, out of the gravitational potential into the IGM. The condition for this is that the speed of the convective wind exceeds the escape speed (Eq. 5.4) or, equivalently, the temperature of the heated gas exceeds the virial temperature (Eq. 5.5).

There are quite a few starburst galaxies fulfilling these criteria. In particular, galaxies with starburst centres such as M 82 (Fig. 5.11) and NGC 253 exhibit clear signs of convective winds transporting thermal and relativistic plasma out into intergalactic space. Low-mass, or dwarf, irregular galaxies possessing large amounts of gas and experiencing episodes of intense star formation are clear candidates for producing galactic winds. In fact, spectroscopic and X-ray observations corroborate the above inferences. The deficiency of synchrotron emission from dwarf galaxies (as compared to massive spirals) is taken as evidence for cosmic rays to be less confined in dwarf galaxies (Fig. 5.12).

Fig. 5.11 The M 81/82 galaxy group mapped at $\lambda = 90$ cm with the WSRT (contours), superimposed onto an optical image (*blueish colour*). The extended synchrotron halo of the starburst galaxy M 82 in the north-west is obvious (Credits: B. Adebahr)

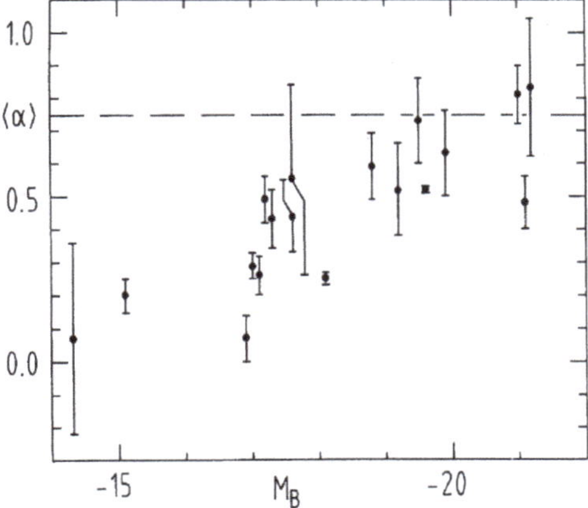

Fig. 5.12 Spectral indices of the radio continuum emission from dwarf galaxies (Klein et al., 1991). The values indicate an increasing fraction of thermal free-free radiation with decreasing optical luminosity

We distinguish 'blow-out' and 'blow-away' scenarios: depending on the energy input and galaxy mass, the (thermal and relativistic) gas is either just blown out of the disk and falls back onto it, which is referred to as a 'galactic fountain', or the material and the magnetic fields escape irreversibly and 'pollute' the ICM/IGM (Mac Low and Ferrara, 1999). This has cosmological significance. In a ΛCDM cosmology, small dark-matter haloes are formed first, with baryonic material subsequently falling into them. Hence, because of the preponderate number of dwarf galaxies, along with their shallow gravitational potentials and their intense star-formation rates (Pop III stars), low-mass galaxies may have (had) a significant role in magnetising their surroundings by injecting relativistic particles and magnetic fields into the ICM/IGM. In fact, (mildy) relativistic particles and magnetic fields are ubiquitously detected in the intra-cluster environment (Sect. 8). In this process, dwarf galaxies must have been competing with AGN. A discussion of this competition must account for the relative powers and duty cycles of activity of dwarf galaxies and AGN (see Sect. 8.7).

Finally, it is interesting to make a quick estimate of the total mass of relativistic particles in a galaxy. This is readily done by assuming equipartition between particles and fields so that the energy densities of the particles (e^-) and magnetic fields are equal:

$$\frac{B^2}{8\,\pi} = n_{rel}\,\gamma\,\eta\,m_e\,c^2\,,$$

where η accounts for the energy contained in the protons (see Chap. 3). This yields the number density of relativistic particles n_{rel}. Multiplying by the likely volume of the relativistic plasma and by the mass of the relativistic protons (which *do* count in this balance), we obtain

$$M_{rel} = n_{rel}\,V\,m_p = \frac{B^2\,m_p}{8\,\pi\,\gamma\,\eta\,m_e\,c^2} \cdot \pi\,R^2\,H,$$

where we have simply assumed a volume V made up by a disk of radius R and thickness H. Assuming $B = 10\,\mu\text{G}$, $\gamma = 3000$, $R = 15\,\text{kpc}$, and $H = 500\,\text{pc}$, we obtain the (perhaps stunningly) tiny amount of $M_{rel} = 0.4\,\text{M}_\odot$! Note that we have chosen a relatively large magnetic-field strength and a small Lorentz factor. This, of course, is the rest mass of the particles, but even accounting for their relativistic energies, meaning that we would have to multiply the above mass by γ^2, would not make this constituent of the ISM of a galaxy dynamically significant.

References

Beck, R., Fletcher, A., Shukurov, A., Snodin, A., Sokoloff, D.D., Ehle, M., Moss, D., Shoutenkov, V., 2005, Astron. Astroph. **444**, 739

Berkhuijsen, E.M., Horellou, C., Krause, M., Neininger, N., Poezd, A.D., Shukurov, A., Sokoloff, D.D., 1997, Astron. Astroph. **318**, 700

Brown, J.C., Taylor, A.R., 2001, Astroph. J. **563**, L31

Chyży, K.T., Beck, R., Kohle, S., Klein, U., Urbanik, M., 2000, Astron. Astroph. **355**, 128

Chyży, K.T., 2008, Astron. Astroph. **482**, 755

de Jong, T., Klein, U., Wielebinski, R., Wunderlich, E., 1985, Astron. Astroph. **147**, L6

Ehle, M., Beck, R., 1993, Astron. Astroph. **273**, 45

Fitt, A.J., Alexander, P., 1993, MNRAS **261**, 445

Fletcher, A., Berkhuijsen, E.M., Beck, R., Shukurov, A., 2004, Astron. Astroph. **414**, 53

Fletcher, A., 2010, in *The Dynamic Interstellar Medium: A Celebration of the Canadian Galactic Plane Survey* eds. R. Kothes, T.L. Landecker, A.G. Willis, ASP Conference Series **438**, 197

Fletcher, A., Beck, R., Shukurov, A., Berkhuijsen, E.M., Horellou, C., 2011, MNRAS **412**, 2396

Frick, P., Beck, R., Shukurov, A., Sokoloff, D., Ehle, M., Kamphuis, J., 2000, MNRAS **318**, 925

Gaensler, B.M., Haverkorn, M., Staveley-Smith, L., Dickey, J.M., McClure-Griffiths, N.M., Dickel, J.R., Wolleben, M., 2005, Science **307**, 1610

Gräve, R., Beck, R., 1988, Astron. Astroph. **192**, 66

Heesen, V., Krause, M., Beck, R., Dettmar, R-J., 2009, Astron. Astroph. **506**, 1123

Helou, G., Soifer, B.T., Rowan-Robinson, M., 1985, Astroph. J. **298**, 7

Houde, M., Fletcher, A., Beck, R., Hildebrand, R.H., Vaillancourt, J.E., Stil, J.M., 2013, Astroph. J. **766**, 49

Hummel, E., 1986, Astron. Astroph. **160**, L4

Jaffe, T.R., Leahy, J.P., Banday, A.J., Leach, S.M., Lowe, S.R., Wilkinson, A., 2010, MNRAS **401**, 1013

Jansson, R., Farrar, G.R., 2012, Astroph. J. **757**, 14

Kennicutt, R.C. Jr., 1981, Astroph. J. **86**, 1847

Kepley, A.A., Mühle, S., Everett, J., Zweibel, E.G., Wilcots, E.M., Klein, U., 2010, Astroph. J. **712**, 536

Klein, U., Weiland, H., Brinks, E., 1991, Astron. Astroph. **246**, 323

Krause, M., 2009, *Magnetic Fields in the Universe II*, Eds. A. Esquivel et al., Revista Mexicana de Astronomia y Astrofisica **36**, 25.

Krause, M., 2012, in *Magnetic Fields in the Universe III: From Laboratory and Stars to Primordial Structures*, Proc. Int. Conf., eds. M. Soida, K. Otmianowska-Mazur, E.M. de Gouveia Dal Pino & A. Lazarian, 155

Krause, M., Hummel, E., Beck, R., 1989, Astron. Astroph. **217**, 4

Krause, M., Beck, R., Hummel, R., 1989, Astron. Astroph. **217**, 17

Longair, M.S., 2011, High-Energy Astrophysics, Cambr. Univ. Press, Chapt. 8

Mac Low, M.-M. Ferrara, A., 1999, Astoph. J. **513**, 142

Mao, S.A., McClure-Griffiths, N.M., Gaensler, B.M., et al., 2012, Astroph. J. **759**, 25

Mora, S.C., Krause, M., 2013, Astron. Astroph. **560**, A42

Nieten, C., Neininger, N., Guélin, M., Ungerechts, H., Lucas, R., Berkhuijsen, E.M., Beck, R., Wielebinski, R., 2006, Astron. Astroph. **453**, 459

Niklas, S., 1995, PhD Thesis, University of Bonn

Oort, J.H., 1974, in *The Formation and Dynamics of Galaxies*, ed. J.R. Shakeshaft, IAU Symposium **58**, 375

Parker, E.N., 1966, Astroph. J. **145**, 811

Patrikeev, I., Fletcher, A., Stepanov, R., Beck, R., Berkhuijsen, E.M., Frick, P., Horellou, C., 2006, Astron. Astroph. **458**, 441

Sandage, A., Humphreys, R.M., 1980, Astroph. J., **236**, L1. Erratum, Astroph. J. **244**, L35

Soida, M., Beck, R., Urbanik, M., Braine, J., 2002, Astron. Astroph. **394**, 47

Soida, M., Krause, M., Dettmar, R.-J. Urbanik, M., 2011, Astron. Astroph. **531**, 127

Sokoloff, D.D., Shukurov, A., Krause, M., 1992, Astron. Astroph. **264**, 396

Sokoloff, D.D., Bykov, A.A., Shukurov, A., Berkhuijsen, E.M., Beck, R., Poezd, A.D., 1998, MNRAS **299**, 189. Erratum, MNRAS **303**, 207

Tabatabaei, F., Krause, M., Fletcher, A., Beck, R., 2008, Astron. Astroph. **490**, 1005

Wunderlich, E., Wielebinski, R., Klein, U., 1987, Astron. Astroph. Suppl. **69**, 487

Chapter 6
Galactic Dynamos

6.1 Introduction

This chapter introduces theoretical ideas that explain some of the common observed properties of galactic magnetic fields. The aim is to give the reader a solid understanding of the basic ideas of galactic dynamo theory, which will allow them to engage with current research in this active field. To begin, we use qualitative arguments to derive a very simple mathematical model for a dynamo that includes some of the key elements of dynamo theory. Then, after discussing some useful concepts from magnetohydrodynamics and the theory of random variables, we derive an induction equation for the mean magnetic field, going into some detail about how the small-scale properties of turbulent flows can lead to large-scale magnetic fields. A detailed mathematical derivation of the basic, dimensionless mean-field dynamo equations for a thin disc follows; by showing explicitly most of the steps involved and clearly identifying assumptions that are made, we hope that even readers who are not enthusiastic about mathematics can gain an intuative feeling for the theory. The main properties of the derived dynamo equations are illustrated by directly solving some simple cases. In addition we provide short, qualitative descriptions of extensions to the basic model that cannot be explicitly derived in a book such as this. We finish with a look at a different dynamo process that has received less attention than the mean-field dynamo, but is perhaps equally important: the fluctuation dynamo that rapidly generates a random magnetic field.

6.2 Qualitative Overview of Galactic Dynamos

The observations described in Chap. 5 show that all disc galaxies contain magnetic fields. In this chapter we will develop a theoretical framework that describes how galactic magnetic fields can be generated, the dynamo. A dynamo is a mechanism

© Springer International Publishing Switzerland 2015
U. Klein, A. Fletcher, *Galactic and Intergalactic Magnetic Fields*,
Springer Praxis Books, DOI 10.1007/978-3-319-08942-3_6

by which kinetic energy is converted into electro-magnetic energy. Bicycle dynamos and wind-up torches use kinetic energy provided by leg or arm muscles to turn a conducting medium, in this case a coil of wire around a fixed ferromagnet, generating a current in the wire which is used to power a light. The conducting medium in the case of the galactic dynamo is the partially ionised interstellar gas, the kinetic energy source is its motion, which occurs on large scales due to galactic rotation and small scales due to turbulence, and the output is the magnetic field. All disc galaxies rotate differentially, so gas at different radii takes different times to complete an orbit of the galactic centre: the signature of this differential rotation is a constant rotational velocity, shown for a sample of spiral galaxies in Fig. 6.1. Differential rotation creates a large-scale shear in the galactic disc that will stretch magnetic field lines in the azimuthal direction, resulting in a stronger magnetic field.

Radio observations show both a large-scale component of the magnetic field, the regular field that is ordered over kiloparsecs, and simultaneously a random small-scale component. It is tempting to associate these two magnetic field components with the respective components of the gas motion. However, although this is the case for the random magnetic field, which is directly amplified by turbulence, it is not the case for the regular magnetic field. A dynamo powered only by the large-scale galactic rotation does not work (we shall take a closer look at anti-dynamo theorems below). Surprisingly, the small-scale random motion of the gas is necessary to produce order in the magnetic field on much larger scales: the galactic dynamo which produces the regular field is a mechanism which extracts order from chaos!

Two of the most important observational constraints on the mechanism for generating regular magnetic fields in galaxy discs are that the fields are always roughly parallel to the disc-plane (the vertical component of the regular field being weak) and that they always have a spiral pattern (the magnetic pitch angle is therefore not zero). There are no observations of a purely circular regular magnetic-field. In cylindrical coordinates we have $B_r \gg B_z$, $B_\phi \gg B_z$, $B_r \neq 0$, $B_\phi \neq 0$ and $B_r < B_\phi$ and so the spirals appear rather tightly wound as the magnetic pitch angle,

$$p_B = \arctan\left(\frac{B_r}{B_\phi}\right), \tag{6.1}$$

tends to be small. The non-zero magnetic pitch angle rules out the simple winding-up of the field lines by the differential rotation of the disc as the origin of the regular fields. The shear due to differentially rotating discs can lead to an amplification of the field by stretching the field lines. However, the problem with this mechanism is the resulting magnetic-field geometry. An initially uniform seed magnetic field would be transformed to one with field directions that alternate with radius over progressively smaller radial intervals as the field lines are wound up. The radial distance between reversals in the field direction would vary with time as $\Delta r \sim r_0/(\Omega_0 t)$, where, for a flat rotation curve with velocity $v_0 \simeq 200\,\mathrm{km\,s^{-1}}$ between $1\,\mathrm{kpc} < r < 10\,\mathrm{kpc}$, $10^{-15}\,\mathrm{s^{-1}} < \Omega_0 < 10^{-14}\,\mathrm{s^{-1}}$. With $r_0 \simeq 10\,\mathrm{kpc}$ and

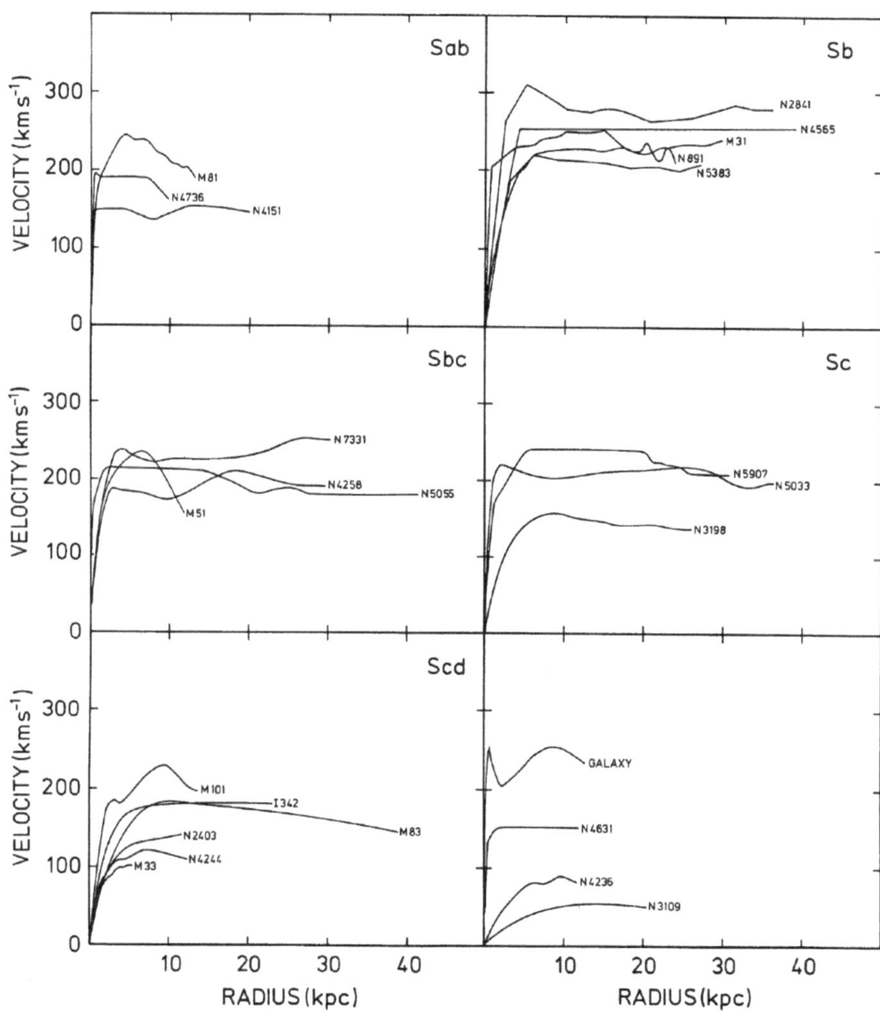

Fig. 6.1 Rotation curves for a sample spiral galaxies, grouped by galaxy class. All the galaxies exhibit non-rigid, or differential, rotation away from their centres. Most have extended radial ranges where the rotation velocity is constant

$\Omega_0 \simeq 10^{-15}\,\mathrm{s}^{-1}$, over the 10^{10} year lifetime of a galaxy $\Delta r \lesssim 100\,\mathrm{pc}$ and $p_\mathrm{B} \simeq 0°$. But neither of these predictions are compatible with the observations.

An alternative is to consider a simple ad-hoc model that can produce a regular magnetic field similar to that observed. In this model, winding up the radial field component B_r will produce an azimuthal component B_ϕ: this is called the Ω-effect, illustrated in Fig. 6.2 with a timescale of τ_Ω. What is now missing is a process to convert a proportion of this B_ϕ back into B_r in order to avoid producing a purely

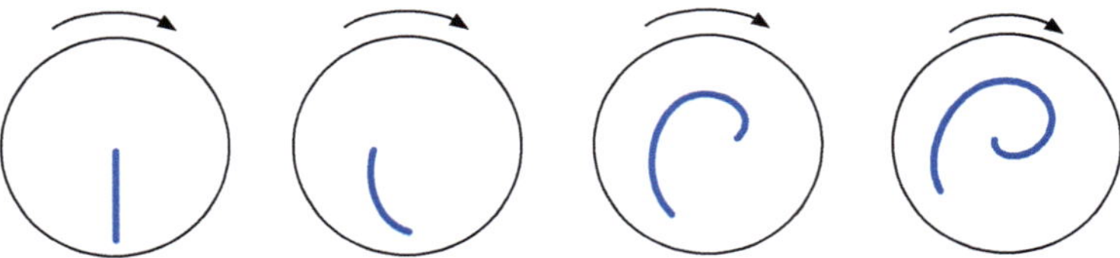

Fig. 6.2 Cartoon showing a physical interpretation of the Ω-effect. As time increases, from left to right, the differential rotation of the disc stretches an initially radial magnetic field line (*blue*) into a field line that has a strong azimuthal (ϕ) component

azimuthal field (i.e. circular field lines, which are not observed). Let us just assume that such a process exists in galactic discs and call the corresponding timescale τ_α. A successful dynamo model must generate electro-magnetic energy faster than it is lost by resistive dissipation into heat, a process controlled by the magnetic diffusivity, which is inversely related to the electrical conductivity. Thus the Ω-effect must produce B_ϕ from B_r faster than B_ϕ is lost due to diffusion, so $\tau_\Omega < \tau_\eta$, and the α-effect must produce B_r from B_ϕ on a timescale that is shorter than the diffusive timescale, so $\tau_\alpha < \tau_\eta$. This model can be written as a coupled pair of ordinary differential equations:

$$\frac{\mathrm{d}B_r}{\mathrm{d}t} = \frac{B_\phi}{\tau_\alpha} - \frac{B_r}{\tau_\eta}, \tag{6.2}$$

$$\frac{\mathrm{d}B_\phi}{\mathrm{d}t} = \frac{B_r}{\tau_\Omega} - \frac{B_\phi}{\tau_\eta}. \tag{6.3}$$

By choosing an initial condition we could solve this system. But all we are concerned with here are the conditions whereby B_r and B_ϕ increase with time, so we can proceed more simply. The solutions will be exponential in time and for a successful dynamo they must increase. Set $B_r = B_{r0}\exp(\gamma t)$ and $B_\phi = B_{\phi 0}\exp(\gamma t)$ where γ is the growth rate for the field: Re $\gamma > 0$ means that the field grows exponentially with time (if γ is complex, then its imaginary part describes oscillations of the growing or decaying solutions). On substitution we obtain

$$\begin{bmatrix} \gamma + 1/\tau_\eta & -1/\tau_\alpha \\ -1/\tau_\Omega & \gamma + 1/\tau_\eta \end{bmatrix} \begin{bmatrix} B_{r0} \\ B_{\phi 0} \end{bmatrix} = 0,$$

for which non-trivial solutions exist when the determinant of the matrix is zero. Thus

$$\gamma = \frac{1}{\sqrt{\tau_\alpha \tau_\Omega}} - \frac{1}{\tau_\eta},$$

and the condition under which the magnetic field will grow, $\gamma > 0$, is

$$\frac{\tau_\alpha}{\tau_\eta}\frac{\tau_\Omega}{\tau_\eta} < 1.$$

This dimensionless condition is stating mathematically what we expect: the timescales for generating the field must be shorter than the diffusive timescale that destroys the field. If we only have a mechanism for generating one field component, for example if $\tau_\alpha \to \infty$, then the solution of Eq. (6.2) has $\gamma < 0$. This simple model exposes the essential features of the mean-field dynamo theory: both B_r and B_ϕ must be generated simultaneously; timescales for generating the field must be shorter than those removing the field; and the condition for growth can be expressed as a critical value of a dimensionless parameter.

6.3 Mean-Field Magnetohydrodynamics

6.3.1 Choice of Units

In astrophysics, the equations of electromagnetism are usually written in Gaussian (or cgs, meaning centimetres, grammes and seconds) units rather than in the SI system. The benefits of working in the cgs system are: no additional constants are necessary (in the SI system the permittivity ϵ_0 and permeability μ_0 of free space are required); no dimensions other than mass M, length L and time T are needed, with electric charge having the dimensions $M^{1/2}L^{3/2}T^{-1}$ (in the SI system a new dimension Q is defined for electric charge); the dimensions of the electric and magnetic fields are the same, both being $M^{1/2}L^{-1/2}T^{-1}$.

As an example of the convenience of working in Gaussian units consider the dimension of B^2, $ML^{-1}T^{-2}$: this is the same as that of energy per unit volume. The magnetic energy density in Gaussian units is simply $B^2/8\pi$ with B measured in Gauss, with no need to remember the value of a constant as is required when working in SI units with B measured in Tesla (the SI constant is μ_0). The energy density of the magnetic field is a quantity which we often wish to calculate, for example to assess how important magnetic fields might be dynamically. For an enlightening and in-depth discussion of systems of units see Chapter 6 of Parker (2007).

6.3.2 Magnetic Induction Equation and the Magnetic Reynolds Number

The fundamental equation that describes the evolution of a magnetic field is the magnetic induction equation. This was derived in Sect. 4.1 and for convenience is written again here, with constant magnetic diffusivity η,

$$\frac{\partial \vec{B}}{\partial t} = \vec{\nabla} \times (\vec{v} \times \vec{B}) + \eta \nabla^2 \vec{B}. \tag{6.4}$$

The first term on the right hand side of Eq. (6.4) describes how the magnetic field interacts with the velocity field of the fluid by advection and stretching of field lines. The second term on the RHS describes the diffusion of the magnetic field. By choosing typical values for the magnetic field strength B_0, velocity U and length scale L of the system, we can write the magnitude of the ratio of the two terms as

$$\frac{|\nabla \times (\vec{v} \times \vec{B})|}{|\eta \nabla^2 \vec{B}|} \sim \frac{(UB_0)/L}{\eta B_0/L^2} = \frac{UL}{\eta} \equiv \mathcal{R}\mathrm{m},$$

where $\mathcal{R}\mathrm{m}$ is called the magnetic Reynolds number: this is a measure of how closely the magnetic field is coupled to the fluid. For high $\mathcal{R}\mathrm{m}$, advection and stretching of the magnetic field by the velocity field of the (partially or fully) ionised fluid is more important than the diffusion of $\vec{B}$. When $\mathcal{R}\mathrm{m} \gg 1$ the magnetic field is said to be "frozen" into the fluid, as discussed in Sect. 4.1. Most astrophysical fluids have a high $\mathcal{R}\mathrm{m}$. When $\mathcal{R}\mathrm{m} \ll 1$ diffusion of the field is more important than advection: this is the usual state of conducting fluids in laboratory experiments.

For spiral galaxies, relevant velocities range between $10\,\mathrm{km\,s^{-1}}$, due to turbulence in the interstellar gas, and $100\,\mathrm{km\,s^{-1}}$, characteristic of the differential rotation. These velocities correspond to length scales of roughly $100\,\mathrm{pc}$ and $10\,\mathrm{kpc}$ respectively, with $\eta = c^2/(4\pi\sigma) \approx 10^{18}\,\mathrm{cm^2\,s^{-1}}$ for the ISM, using Eq. (4.6). Thus $\mathcal{R}\mathrm{m} \approx 10^8$ for the turbulent magnetic field and $\mathcal{R}\mathrm{m} \approx 10^{11}$ for the regular magnetic field. Such a large $\mathcal{R}\mathrm{m}$ means that we can think of the magnetic field lines being dragged around by the gas. Magnetic diffusivity becomes important on much smaller scales. However, topological re-organisation of the magnetic field lines can only take place in the MHD model when the diffusivity is not zero: $\eta \neq 0$ allows magnetic lines to reconfigure themselves in different arrangements, for example to form new closed loops of field, which is not possible when $\eta = 0$.

As well as Eq. (6.4), a full dynamical theory of dynamo action requires additional equations that describe the evolution of the fluid's velocity field (the mass conservation, or continuity, equation and the momentum conservation, or Navier-Stokes, equation) and its thermodynamic state (its energy evolution). Both the Navier-Stokes and energy equations will have extra terms, compared to their usual hydrodynamic forms. In the case of the fluid momentum equation this term describes the Lorentz force $\vec{j} \times \vec{B} \propto \nabla \times \vec{B} \times \vec{B}$[1] produced by the magnetic field on the ionised fluid and for the energy equation a term describing Ohmic heating is

---

[1] Note that the current density $\vec{j}$ is not a variable in the induction equation. Along with the electric field $\vec{E}$, it is eliminated in favour of $\vec{B}$ and $\vec{v}$ which are the natural variables to use in describing astrophysical magnetic fields. If necessary $\vec{j}$ and $\vec{E}$ can be recovered from $\vec{v}$ and $\vec{B}$ using Maxwell's equations and Ohm's Law.

required. A suitable set of boundary and initial conditions are needed to provide a complete mathematical description of the problem. If one writes out the full set of equations involved, it is immediately clear that a comprehensive, dynamical description of magnetic field evolution is a formidable problem that can only be tackled computationally (and even then many compromises and assumptions are required).

Here we will restrict ourselves to discussing a kinematic dynamo, that is one where the velocity field $\vec{v}$ is prescribed and is not altered by the magnetic field. The kinematic regime is one which we expect to occur during the early stages of dynamo action. However, once the magnetic energy density approaches the energy density of the fluid, it is not plausible to ignore the Lorentz force. A simple algebraic parameterisation to model this back-reaction, and so to allow the magnetic field to reach an equilibrium state rather than grow continually, is discussed briefly at the end of this chapter. Note that all dynamos require the existence of an initial seed magnetic field, which can be extremely weak but never zero, on which to act.

6.3.3 Anti-dynamo Theorems

In the early twentieth century the origin of the dipolar magnetic fields of the Earth and Sun were pressing problems. It was known that ferromagnetism could not be responsible due to the high temperatures involved and that both bodies contained conducting fluids (the molten iron outer-core of the Earth and the ionised gas of the Sun) with a strong rotational component to their flow. It was a natural question to ask whether these large-scale flows were responsible for the observed magnetic fields. By analogy with electro-magnetism experiments, a current inside the Earth parallel to the equator could maintain a dipolar magnetic field. The question was: could the currents generated in a rotating conducting fluid do the same job?

The simplest model to consider was the case of a steady, axisymmetric velocity field (the conducting fluid moving around the axis of rotation) supporting a steady, axisymmetric magnetic field (ignoring the slight offset between the rotational and magnetic axes). In 1934 Cowling proved that such a model cannot work. It is necessary to relax one or more of the restrictions in order to obtain a working dynamo. Parker (1955) proposed a way to solve this problem by including small-scale components of the flow. All astrophysical flows are turbulent due to their high Reynolds number $\mathcal{Re} \equiv UL/\nu$, where U is the velocity, L the length scale and $\nu \sim 10^{7} n^{-1} T^{5/2}\,\mathrm{cm^2\,s^{-1}}$ the kinematic viscosity for a fully ionised hydrogen plasma of temperature T and number density n. Parker's insight was that a combination of density stratification, which results in the buoyancy of fluid elements that are lighter than their surroundings, and large-scale rotation, producing a Coriolis force, will lead to rising turbulent eddies with a common sign of vorticity. Thus on average the random turbulent motions can have a large-scale effect on the generated magnetic field. All of the necessary requirements for this effect are present in the interstellar gas in galactic discs. This is the basic physical picture underlying the α-effect that was used in the qualitative picture discussed above.

Cowling's theorem was the first of several anti-dynamo theorems, each of which puts constraints on the type of magnetic fields that can be supported by different velocity fields. Others show that neither two-dimensional flows nor purely toroidal flows can maintain a dynamo. For a detailed discussion of anti-dynamo theorems see Jones (2008).

6.3.4 Reynolds Averaging

The magnetic induction equation describes the evolution of the magnetic field on all scales. Observations indicate that galactic magnetic fields can be roughly separated into two components, a regular field that is structured on scales comparable to the galactic radius of about 10 kpc and a random field that has a correlation scale of only 10–100 pc. The origin of the regular magnetic field has always attracted most attention and this is where we will focus our attention. To develop an understanding about how the magnetic fields evolve we can feel justified in making several simplifications. First, we shall average Eq. (6.4). This will split it into two parts, describing the evolution of the large-scale and small-scale fields separately. The averaging allows us to treat the random fluctuations in the velocity, due to turbulence, and the random fluctuations in the magnetic field via their statistical properties. The averaging process that we apply follows particular rules, named after Osborne Reynolds, the pioneering ninetieth century fluid dynamicist.

The Reynolds rules for averaging random quantities are as follows. Let $x(r)$ and $y(r)$ be two random variables which are functions of the independent variable r and let a be some scalar. The averaging procedure denoted $\langle \ldots \rangle$ should have the following properties:

$$\langle x + y \rangle = \langle x \rangle + \langle y \rangle, \tag{6.5}$$

$$\langle ax \rangle = a \langle x \rangle, \tag{6.6}$$

$$\langle \langle x \rangle y \rangle = \langle x \rangle \langle y \rangle \implies \langle \langle x \rangle \rangle = \langle x \rangle, \tag{6.7}$$

$$\langle \mathrm{d}x/\mathrm{d}r \rangle = \mathrm{d}\langle x \rangle/\mathrm{d}r, \tag{6.8}$$

$$\left\langle \int x\,\mathrm{d}r \right\rangle = \int \langle x \rangle \mathrm{d}r. \tag{6.9}$$

The appropriate averaging procedure used in the theory of random variables is to take the ensemble-average of many independent realisations of the system. In laboratory fluid dynamics one could conduct multiple experiments with identical configurations and average the measured quantities. This allows us to write a random variable, measured in a particular experiment, as the sum of an ensemble-average and a fluctuation which is specific to that particular realisation,

$$x = \langle x \rangle + x'. \tag{6.10}$$

Averaging Eq. (6.10) and applying the rules given by Eqs. (6.5) and (6.7),

$$\langle x' \rangle = 0. \tag{6.11}$$

Since it is not possible to carry out ensemble-averaging for real galaxies (although in principle we can do this in the case of numerical simulations), in practice we assume that spatial averages, which we can measure, are equivalent to the ensemble-averages which are used to develop the theory of random flows.

6.3.5 Mean-Field Magnetic Induction Equation

We are now ready to derive the induction equation for the mean component of the magnetic field. First, use Eq. (6.10) to write $\vec{B}$ and $\vec{v}$ as the sum of a mean and fluctuating part, $\vec{B} = \langle \vec{B} \rangle + \vec{B}'$ and $\vec{v} = \langle \vec{v} \rangle + \vec{v}'$. Equation (6.4) then becomes

$$\frac{\partial(\langle \vec{B} \rangle + \vec{B}')}{\partial t} = \nabla \times [(\langle \vec{v} \rangle + \vec{v}') \times (\langle \vec{B} \rangle + \vec{B}')] + \eta \nabla^2(\langle \vec{B} \rangle + \vec{B}'). \tag{6.12}$$

Next we will average each of the terms in Eq. (6.12). To illustrate this procedure, let us follow step-by-step what happens when taking the average of the term $\nabla \times (\langle \vec{v} \rangle \times \vec{B}')$. Equation (6.8) allows us to write

$$\langle \nabla \times (\langle \vec{v} \rangle \times \vec{B}') \rangle = \nabla \times \langle \langle \vec{v} \rangle \times \vec{B}' \rangle,$$

then use Eq. (6.7) to write

$$\nabla \times \langle \langle \vec{v} \rangle \times \vec{B}' \rangle = \nabla \times (\langle \vec{v} \rangle \times \langle \vec{B}' \rangle),$$

and as $\langle \vec{B}' \rangle = 0$ by Eq. (6.11) we find that

$$\langle \nabla \times (\langle \vec{v} \rangle \times \vec{B}') \rangle = 0.$$

Taking the average of each term in Eq. (6.12) and applying the properties of the averaging procedure given by Eqs. (6.5)–(6.11), we are left with

$$\frac{\partial \langle \vec{B} \rangle}{\partial t} = \nabla \times (\langle \vec{v} \rangle \times \langle \vec{B} \rangle) + \nabla \times \langle \vec{v}' \times \vec{B}' \rangle + \eta \nabla^2 \langle \vec{B} \rangle, \tag{6.13}$$

where only one term containing the fluctuations $\vec{B}'$ and $\vec{v}'$ remains.

Note that

$$\langle \vec{v}' \times \vec{B}' \rangle \neq \langle \vec{v}' \rangle \times \langle \vec{B}' \rangle$$

if the fluctuations are correlated; for example, if most realisations of the system produce the same sign (and a similar magnitude) for the quantity $\vec{v}' \times \vec{B}'$ then taking the ensemble-average of a large number of realisations will not result in mutual cancellation of these terms. This is indeed what we expect, because both ways in which $\vec{B}'$ can be produced depend on the statistical properties of $\vec{v}'$, either tangling of the mean-field $\langle \vec{B} \rangle$ by $\vec{v}'$, or the direct generation of $\vec{B}'$ by $\vec{v}'$ via the fluctuation dynamo mechanism (see Sect. 6.5).

Equation (6.13) says something remarkable which we should not let the mathematics hide: the evolution of the large-scale magnetic field $\langle \vec{B} \rangle$ depends on both, the large-scale velocity $\langle \vec{v} \rangle$ which is essentially the differential rotation of the gaseous disc, and the small-scale turbulence $\vec{v}'$. Order is produced from the chaotic random velocities! The book by Zel'dovich et al. (1990) is recommended for more on this and other unexpected properties of random quantities.

6.3.6   *Alpha Effect and Turbulent Diffusivity*

Equation (6.13) describes the evolution of the mean magnetic field $\langle \vec{B} \rangle$ in terms of the mean velocity field $\langle \vec{v} \rangle$ and the averaged product of the fluctuations in $\vec{v}$ and $\vec{B}$. But it is difficult to know how to deal with the term $\mathcal{E} \equiv \langle \vec{v}' \times \vec{B}' \rangle$. This is called the mean electromotive force (EMF), as it is an extra source of current in the mean-field version of Ohm's Law,

$$\langle \vec{J} \rangle = \sigma \left[\langle \vec{E} \rangle + \frac{1}{c}(\langle \vec{v} \rangle \times \langle \vec{B} \rangle + \mathcal{E}) \right],$$

obtained from Eq. (4.4) by changing variables using Eq. (6.10) and then averaging. Re-writing this term using only the mean magnetic field $\langle \vec{B} \rangle$ and statistical properties of the turbulent velocity $\vec{v}'$ would be advantageous. The development of methods to do this, first by Parker (1955) using physical arguments and later by Steenbeck et al. (1966) following a systematic mathematical approach, resulted in the theory of the turbulent mean-field dynamo, a theory that is relevant to the generation of magnetic fields in many astrophysical objects such as the Earth, Sun, stars and accretion discs, as well as galaxies.

Since the pioneering work of Parker (1955) and Steenbeck et al. (1966) there have been vigorous discussions on the validity of the different physical and mathematical arguments proposed to deal with the $\langle \vec{v}' \times \vec{B}' \rangle$ term in Eq. (6.13). All aim to expand the mean EMF, retaining only terms linear in $\langle \vec{B} \rangle$ and its derivatives, where the coefficients of the retained terms are averaged properties of the turbulent velocity

$\vec{v}'$. The description of turbulence is greatly simplified when it is homogeneous and isotropic, so this assumption is often made.

For illustration we shall present one approach, called the first order smoothing approximation (FOSA), following the derivation in Choudhuri (1998, Chapter 16.5). For the sake of simplicity we will work in a Cartesian reference frame (x, y, z). In the FOSA approach the crucial assumption is that $\vec{B}' \ll \langle \vec{B} \rangle$, which is generally not the case in the later stages of magnetic field evolution in galaxies, but may be applicable at early times. By subtracting Eq. (6.13) from Eq. (6.12) we obtain an induction equation describing the evolution of the fluctuating magnetic field:

$$\frac{\partial \vec{B}'}{\partial t} = \nabla \times (\vec{v}' \times \langle \vec{B} \rangle + \langle \vec{v} \rangle \times \vec{B}' + \vec{v}' \times \vec{B}' - \langle \vec{v}' \times \vec{B}' \rangle) + \eta \nabla^2 \vec{B}'. \qquad (6.14)$$

Now consider what happens to the mean field $\langle \vec{B} \rangle$ at a fixed point, over the time it takes for the local velocity $\vec{v}$ to change appreciably due to $\vec{v}'$, which is called the correlation time τ. Initially assume that $\vec{v} = \vec{v}_0 \approx \langle \vec{v} \rangle$ and $\vec{v}' = 0$, so the evolution of $\vec{B}'$ is governed by the 2nd term on the RHS of Eq. (6.14). But if $\vec{B}' \approx 0$ initially, then there is no generation of new $\vec{B}'$. Once the time interval exceeds τ though, $|\vec{v} - \vec{v}_0| \sim |\vec{v}'| > 0$ and the 1st term on the RHS of Eq. (6.14) will generate $\vec{B}'$ from $\langle \vec{B} \rangle$: the random, turbulent velocities will start to tangle up the existing mean magnetic field. So, neglecting the terms containing $\vec{B}'$ on the RHS of Eq. (6.14), we have an approximate equation for the evolution of $\vec{B}'$ in terms of $\vec{v}'$ and $\langle \vec{B} \rangle$,

$$\frac{\partial \vec{B}'}{\partial t} \approx \nabla \times \vec{v}' \times \langle \vec{B} \rangle = (\langle \vec{B} \rangle \cdot \nabla)\vec{v}' - (\vec{v}' \cdot \nabla)\langle \vec{B} \rangle, \qquad (6.15)$$

where incompressibility of the fluid, $\nabla \cdot \vec{v} = 0$, has been assumed and we recall that $\nabla \cdot \vec{B} = 0$. Making the approximation $\partial \vec{B}'/\partial t \sim \vec{B}'/\tau$ we obtain from Eq. (6.15)

$$\vec{v}' \times \vec{B}' = \tau[\vec{v}' \times (\langle \vec{B} \rangle \cdot \nabla)\vec{v}' - \vec{v}' \times (\vec{v}' \cdot \nabla)\langle \vec{B} \rangle]. \qquad (6.16)$$

Then we take the ensemble-average to obtain $\langle \vec{v}' \times \vec{B}' \rangle$ expressed in terms involving $\vec{v}'$ and $\langle \vec{B} \rangle$ only. To keep the equations relatively compact, it is necessary to proceed using tensor notation, with repeated subscripts indicating a summation over all axes. The i-th component of the two terms in $\langle \vec{v}' \times \vec{B}' \rangle$ are then

$$\tau[\langle \vec{v}' \times (\langle \vec{B} \rangle \cdot \nabla)\vec{v}' \rangle]_i = \epsilon_{ijk}\langle v'_j \langle B_l \rangle \frac{\partial v'_k}{\partial x_l} \rangle \tau = \alpha_{ij}\langle B_j \rangle, \quad \alpha_{ij} \equiv \epsilon_{ilk}\langle v'_l \frac{\partial v'_k}{\partial x_j} \rangle \tau,$$

and

$$\tau[\langle \vec{v}' \times (\vec{v}' \cdot \nabla)\langle \vec{B} \rangle \rangle]_i = \epsilon_{ijk}\langle v'_j v'_l \frac{\partial \langle B_k \rangle}{\partial x_l} \rangle \tau = \eta_{ijk}\frac{\partial \langle B_k \rangle}{\partial x_j}, \quad \eta_{ijk} \equiv \epsilon_{ilk}\langle v'_l v'_j \rangle \tau,$$

where summation indices have been relabelled as necessary.

If the turbulence is isotropic, that is if its properties such as α_{ij} and η_{ijk} are invariant under rotation, then the two tensors become

$$\alpha_{ij} = \alpha\delta_{ij}, \quad \alpha = -\frac{1}{3}\langle \vec{v}' \cdot (\nabla \times \vec{v}')\rangle\tau, \tag{6.17}$$

and

$$\eta_{ijk} = \eta_T\epsilon_{ijk}, \quad \eta_T = \frac{1}{3}\langle \vec{v}' \cdot \vec{v}'\rangle\tau. \tag{6.18}$$

The quantity $\vec{v}' \cdot (\nabla \times \vec{v}')$ is called the helicity of the turbulent velocity.

Thus, the expansion of the mean EMF term is (for its i-th component),

$$\langle \vec{v}' \times \vec{B}'\rangle_i \approx \alpha_{ij}\langle B_j\rangle - \eta_{ijk}\frac{\partial\langle B_k\rangle}{\partial x_j}.$$

The α-effect in Eq. (6.17) is a mathematical representation of Parker's insight into how stratification and rotation will combine to produce a non-zero average helicity of the turbulence. The handedness of the twisting will be the same for all of the rising or falling turbulent eddies on the same side of the mid-plane, if we assume that the Coriolis force is the cause of the twisting, and so the small-scale turbulence produces a large-scale effect. The sign of the helicity is different above and below the mid-plane, as the vertical component of the velocity of the buoyant eddies have opposite signs, so α will be an odd function in z with $\alpha(-z) = -\alpha(z)$ and $\alpha(0) = 0$ at the mid-plane. Figure 6.3 shows a schematic representation of the α-effect for a magnetic field line being moved by a single eddy.

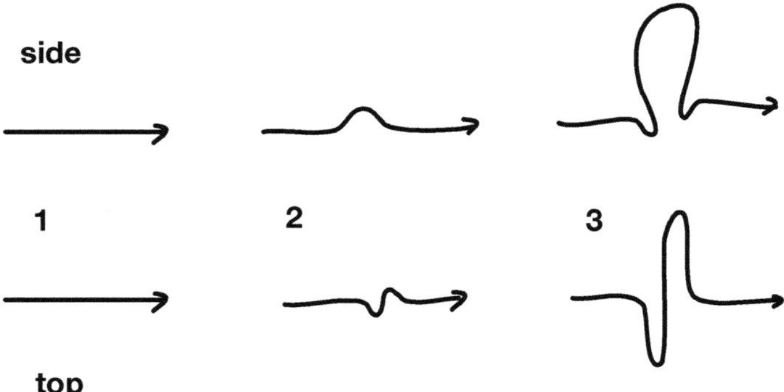

Fig. 6.3 Cartoon showing a physical interpretation of the α-effect. An initially straight magnetic field line (*1*), begins to be lifted in the vertical direction by a parcel of buoyant gas (*2*), and is twisted as it rises by the Coriolis force due to the whole system rotating (*3*). The *top row* shows a side-on view of each step, the *bottom row* a top-down view. If the initial magnetic field line is directed in the azimuthal ϕ direction, the end result is a field line with both r and ϕ components

If the helicity in Eq. (6.17) is due to buoyant turbulent cells being twisted by the Coriolis force, a useful estimate for the order of magnitude of α is

$$\alpha \sim \frac{l^2 \Omega}{h}, \tag{6.19}$$

where l is the correlation length of the turbulence, Ω the angular velocity of the rotating disc and h the disc scale-height. On the other hand, since α is a component of the turbulent velocity it should never exceed $\vec{v}'$.

In Sect. 6.4 we shall see that the α-effect becomes a generating term in the dynamo equations, allowing for the conversion of azimuthal field B_ϕ into radial field B_r. Since the flow is turbulent, i.e. neither steady nor axisymmetric, the constraints of Cowling's anti-dynamo theorem are removed and a working dynamo model can be obtained that can generate an axisymmetric magnetic field.

Using Eqs. (6.17) and (6.18) we can re-write the mean electromotive force term in Eq. (6.13) as

$$\nabla \times \langle \vec{v}' \times \vec{B}' \rangle \approx \nabla \times (\alpha \langle \vec{B} \rangle - \eta_T \nabla \times \langle \vec{B} \rangle) = \nabla \times \alpha \langle \vec{B} \rangle + \eta_T \nabla^2 \langle \vec{B} \rangle, \tag{6.20}$$

where any spatial variation in η_T has been ignored in the last step. This shows the rôle of η_T, defined in Eq. (6.18), to be that of a diffusion coefficient. The diffusion of $\langle \vec{B} \rangle$ is therefore due to both, η_T the turbulent cascade moving the mean magnetic field rapidly to small scales, and η the slower effect of resistivity. A rough estimate for η_T is

$$\eta_T = \frac{\tau v^2}{3} \sim \frac{l}{v}\frac{v^2}{3} = \frac{lv}{3} \sim 10^{26}\,\mathrm{cm}^2\,\mathrm{s}^{-1}, \tag{6.21}$$

for turbulent velocities $v = 10\,\mathrm{km\,s}^{-1}$ with a correlation length of $l = 100\,\mathrm{pc}$, showing that the turbulent diffusivity is a much stronger dissipative effect than the standard resistive diffusion with $\eta \approx 10^{18}\,\mathrm{cm}^2\,\mathrm{s}^{-1}$. An estimate of the diffusion time scale, the time it takes the magnetic field to dissipate, for a magnetic field in a disc with a thickness of $H = 1\,\mathrm{kpc}$ is then $\tau_B \sim H^2/4\eta_T \approx 10^8$ year, which is a small fraction of the lifetime of a galaxy. Compare this timescale with the much longer timescale for field decay obtained in Sect. 4.2.2 using only Ohmic resistivity.

Several other approaches to dealing with the troublesome $\langle \vec{v}' \times \vec{B}' \rangle$ term in the mean-field induction equation have been developed and there is still considerable debate about the validity of the mean-field approach, in particular regarding the nature and even existence of the α-effect. An interested reader will find detailed derivations based on the assumption of a short correlation time of the turbulence, rather than the FOSA assumption of $\vec{B}' \ll \langle \vec{B} \rangle$, in Ruzmaikin et al. (1988) and Kulsrud (2005), and a discussion of many of the different schemes in Brandenburg and Subramanian (2005).

6.4 Mean-Field Dynamo

We can now investigate the mean-field galactic dynamo more quantitatively. Combine Eqs. (6.13) and (6.20) to obtain

$$\frac{\partial \vec{B}}{\partial t} = \nabla \times (\vec{v} \times \vec{B}) + \nabla \times \alpha \vec{B} + \eta_T \nabla^2 \vec{B}, \tag{6.22}$$

where we have dropped the averaging notation $\langle \ldots \rangle$ since $\vec{v}$ and $\vec{B}$ are now mean fields and we only retain η_T, which is assumed to be constant in space and time, as $\eta \ll \eta_T$. Working in a cylindrical coordinate system (r, ϕ, z), let us assume that

$$\vec{v} = [0, \Omega(r)r, 0] \tag{6.23}$$

and that $\vec{B} = \vec{B}(r, z, t)$; in other words we have an axisymmetric galaxy disc, so $\partial/\partial\phi = 0$. We shall also assume that $\alpha = \alpha(z)$.

The Laplacian in the diffusion term of Eq. (6.22) is

$$\nabla^2 \vec{B} \equiv \nabla(\nabla \cdot \vec{B}) - \nabla \times (\nabla \times \vec{B}), \tag{6.24}$$

so using the condition $\nabla \cdot \vec{B} = 0$ we obtain, for axisymmetric $\vec{B}$,

$$\begin{aligned}
\nabla^2 \vec{B} &= -\nabla \times (\nabla \times \vec{B}) \\
&= \left[\frac{\partial^2 B_r}{\partial z^2} - \frac{\partial^2 B_z}{\partial z \partial r} \right] \vec{\hat{r}} \\
&\quad + \left[\frac{\partial^2 B_\phi}{\partial z^2} + \frac{\partial}{\partial r} \left(\frac{1}{r} \frac{\partial}{\partial r} (r B_\phi) \right) \right] \vec{\hat{\phi}} \\
&\quad + \left[\frac{1}{r} \frac{\partial}{\partial r} \left(r \frac{\partial B_z}{\partial r} \right) - \frac{1}{r} \frac{\partial}{\partial r} \left(r \frac{\partial B_r}{\partial z} \right) \right] \vec{\hat{z}}.
\end{aligned} \tag{6.25}$$

Then use the divergence-free magnetic field condition again,

$$\nabla \cdot \vec{B} = \frac{1}{r} \frac{\partial}{\partial r} (r B_r) + \frac{1}{r} \frac{\partial B_\phi}{\partial \phi} + \frac{\partial B_z}{\partial z} = 0, \tag{6.26}$$

to obtain, for axisymmetric $\vec{B}$,

$$\frac{1}{r} \frac{\partial}{\partial r} (r B_r) = -\frac{\partial B_z}{\partial z}, \tag{6.27}$$

which gives

$$-\frac{\partial^2 B_z}{\partial z \partial r} = -\frac{\partial}{\partial r} \frac{\partial B_z}{\partial z} = \frac{\partial}{\partial r} \left[\frac{1}{r} \frac{\partial}{\partial r} (r B_r) \right],$$

and

$$-\frac{1}{r}\frac{\partial}{\partial r}\left(r\frac{\partial B_r}{\partial z}\right) = -\frac{\partial}{\partial z}\left[\frac{1}{r}\frac{\partial}{\partial r}(rB_r)\right] = \frac{\partial^2 B_z}{\partial z^2}.$$

Expanding the terms involving $\vec{v}$ and α in Eq. (6.22) gives the following three equations describing the evolution of the three components of the magnetic field:

$$\frac{\partial B_r}{\partial t} = -\frac{\partial}{\partial z}(\alpha B_\phi) + \eta_T\frac{\partial^2 B_r}{\partial z^2} + \eta_T\frac{\partial}{\partial r}\left[\frac{1}{r}\frac{\partial}{\partial r}(rB_r)\right], \tag{6.28}$$

$$\frac{\partial B_\phi}{\partial t} = \frac{\partial}{\partial z}(\Omega\, rB_z) + \frac{\partial}{\partial r}(\Omega\, rB_r) + \frac{\partial}{\partial z}(\alpha B_r) - \frac{\partial}{\partial r}(\alpha B_z)$$

$$+\eta_T\frac{\partial^2 B_\phi}{\partial z^2} + \eta_T\frac{\partial}{\partial r}\left[\frac{1}{r}\frac{\partial}{\partial r}(rB_\phi)\right], \tag{6.29}$$

$$\frac{\partial B_z}{\partial t} = \frac{1}{r}\frac{\partial}{\partial r}(r\alpha B_\phi) + \eta_T\frac{\partial^2 B_z}{\partial z^2} + \eta_T\frac{1}{r}\frac{\partial}{\partial r}\left(r\frac{\partial B_z}{\partial r}\right). \tag{6.30}$$

The workings of the dynamo are starting to become clear. Each component of the magnetic field is diluted by diffusion, both vertically and radially, and each has at least one source term. B_r and B_z are created from B_ϕ due to the helicity of the turbulent velocity, encapsulated in α, and B_ϕ is created by shearing of B_r and B_z in the differential rotation and by the α-effect twisting these two components.

Now let us seek solutions $\vec{B}(z, t)$ at a fixed radius r by assuming that the solutions for the magnetic field components can be separated in r and z as

$$\vec{B}(r, z, t) = Q(r)\vec{B}(z, t), \tag{6.31}$$

that is into a local evolution of $\vec{B}$ at fixed r and a global behaviour $Q(r)$. This is motivated by the small aspect ratio of galactic discs, where $h/R_0 \ll 1$ for disc scale-height h and disc radius R_0, so that $\partial/\partial z \gg \partial/\partial r$. The linearity in B_i and Q of the equation system gives a time dependence in the solutions of the form $B_i(z, t) \propto \exp(\gamma t)$, where i represents the three components of the field (r, ϕ, z), and $B_i(r, z, t) \propto \exp \Gamma t$, where due to the shorter diffusion time in z compared to that in r we expect $\gamma > \Gamma$. We shall first concentrate on developing the equations for $B_i(z, t)$ and later derive an appropriate evolution equation for $B_i(r, z, t)$.

The divergence-free condition becomes

$$\nabla \cdot \vec{B} = \frac{B_r}{r} + \frac{\partial B_z}{\partial z} = 0,$$

which is used to simplify the terms in Eq. (6.29) involving the rotation,

$$\frac{\partial}{\partial z}[\Omega(r)\, rB_z(z, t)] + \frac{\partial}{\partial r}[\Omega(r)\, rB_r(z, t)] = -\Omega B_r + \Omega B_r + B_r\, r\frac{\partial\Omega}{\partial r}.$$

Some other terms also simplify,

$$\frac{\partial}{\partial r}[\alpha(z)B_z(z,t)] = 0,$$

$$\frac{1}{r}\frac{\partial}{\partial r}[r\alpha(z)B_\phi(z,t)] = \frac{\alpha B_\phi}{r},$$

$$\eta_T\frac{\partial}{\partial r}\left\{\frac{1}{r}\frac{\partial}{\partial r}[rB_r(z,t)]\right\} = -\eta_T\frac{B_r}{r^2},$$

with a similar equation for B_ϕ, and

$$\eta_T\frac{1}{r}\frac{\partial}{\partial r}\left(r\frac{\partial B_z}{\partial r}\right) = 0.$$

So we obtain

$$\frac{\partial B_r}{\partial t} = -\frac{\partial}{\partial z}(\alpha B_\phi) + \eta_T\frac{\partial^2 B_r}{\partial z^2} - \eta_T\frac{B_r}{r^2}, \tag{6.32}$$

$$\frac{\partial B_\phi}{\partial t} = B_r\,r\,\frac{d\Omega}{dr} + \frac{\partial}{\partial z}(\alpha B_r) + \eta_T\frac{\partial^2 B_\phi}{\partial z^2} - \eta_T\frac{B_\phi}{r^2}, \tag{6.33}$$

$$\frac{\partial B_z}{\partial t} = \frac{\alpha B_\phi}{r} + \eta_T\frac{\partial^2 B_z}{\partial z^2}. \tag{6.34}$$

6.4.1 Dimensionless Equations

Equations (6.32)–(6.34) can be further simplified by introducing dimensionless variables. This procedure also has the advantage of isolating the dimensionless combinations of parameters that determine the properties of the solutions. Define new dimensionless variables, denoted using an asterisk $*$: $B_i = B_0 B_i^*$, $r = R_0 r^*$, $z = hz^*$, $t = (h^2/\eta_T)t^*$, $\alpha = \alpha_0\alpha^*$, $\Omega = \Omega_0\Omega^*$, $\partial/\partial r = (1/R_0)\partial/\partial r^*$, $\partial/\partial z = (1/h)\partial/\partial z^*$, $\partial^2/\partial z^2 = (1/h^2)\partial^2/\partial z^{*2}$ and $\partial/\partial t = (\eta_T/h^2)\partial/\partial t^*$, where h is the disc scale-height, α_0 and Ω_0 are the typical magnitudes of the α-effect and angular velocity, R_0 is the disc radius and B_0 is an arbitrary scaling for the magnetic field. Substituting these new variables into Eqs. (6.32)–(6.34) gives

$$\frac{\partial B_r^*}{\partial t^*} = -\frac{h\alpha_0}{\eta_T}\frac{\partial}{\partial z^*}(\alpha^* B_\phi^*) + \frac{\partial^2 B_r^*}{\partial z^{*2}} - \left(\frac{h}{R_0}\right)^2\frac{B_r^*}{r^{*2}},$$

$$\frac{\partial B_\phi^*}{\partial t^*} = \frac{h^2\Omega_0}{\eta_T}B_r^*\,r^*\,\frac{d\Omega^*}{dr^*} + \frac{h\alpha_0}{\eta_T}\frac{\partial}{\partial z^*}(\alpha^* B_r^*) + \frac{\partial^2 B_\phi^*}{\partial z^{*2}} - \left(\frac{h}{R_0}\right)^2\frac{B_\phi^*}{r^{*2}},$$

$$\frac{\partial B_z^*}{\partial t^*} = \left(\frac{h}{R_0}\right)\frac{h\alpha_0}{\eta_T}\frac{\alpha^*}{r^*}B_\phi^* + \frac{\partial^2 B_z^*}{\partial z^{*2}}.$$

Dropping the * since all variables are now dimensionless, neglecting the terms which contain $h/R_0 \ll 1$ to the first power and higher, and defining

$$\mathcal{R}_\alpha \equiv \frac{\alpha_0 h}{\eta_T}, \quad \mathcal{R}_\Omega \equiv \frac{\Omega_0 h^2}{\eta_T}, \tag{6.35}$$

we are left with

$$\frac{\partial B_r}{\partial t} = -\mathcal{R}_\alpha \frac{\partial}{\partial z}(\alpha B_\phi) + \frac{\partial^2 B_r}{\partial z^2}, \tag{6.36}$$

$$\frac{\partial B_\phi}{\partial t} = \mathcal{R}_\Omega B_r \, r \frac{d\Omega}{dr} + \mathcal{R}_\alpha \frac{\partial}{\partial z}(\alpha B_r) + \frac{\partial^2 B_\phi}{\partial z^2}, \tag{6.37}$$

$$\frac{\partial B_z}{\partial t} = \frac{\partial^2 B_z}{\partial z^2}. \tag{6.38}$$

These are the basic mean-field galactic dynamo equations in an axisymmetric system.

Ignoring the term $(h/R_0)(h\alpha_0/\eta_T)B_\phi^*$, on the grounds that $h/R_0 \ll 1$ whereas the remaining terms in the dimensionless system of equations are of order 1 (we expect $\mathcal{R}_\alpha \geq 1$ and $\mathcal{R}_\Omega \geq 1$), means that Eq. (6.38) is decoupled from Eqs. (6.36) and (6.37). This makes the search for solutions simpler, but the justification for dropping a term that is only first order in h/R_0 is weaker than for ignoring higher order terms. Equation (6.38) is a diffusion equation with decaying solutions of the form $B_z \propto \exp(\gamma t)$ with $\gamma < 0$, so we only need to solve the coupled pair of differential equations describing the evolution of B_r and B_ϕ. This model is compatible with the absence of B_z in observations of galactic discs, or at least its weakness in comparison to the observed B_r and B_ϕ. If necessary, B_z could be constructed from the solutions to Eqs. (6.36) and (6.37) using $\nabla \cdot \vec{B} = 0$ but, particularly in the case of numerical studies, it may be preferable to retain the term that is first-order in h/R_0 and directly obtain B_z.

$\mathcal{R}_\alpha$ and $\mathcal{R}_\Omega$ are Reynolds numbers for the α- and Ω-effects. They compare the efficiency of magnetic field generation to its loss due to diffusion, and play a similar rôle to the dimensionless timescales τ_α/τ_η and τ_Ω/τ_η introduced qualitatively in Sect. 6.2.

When the α-effect is less efficient at producing B_ϕ from B_r than the differential rotation, that is when $\mathcal{R}_\alpha < \mathcal{R}_\Omega$ (which is the usual case in differentially rotating spiral and barred galaxies), we can drop the $\mathcal{R}_\alpha$ term in Eq. (6.37) and the equations describe an $\alpha\Omega$-dynamo. In dwarf galaxies, where differential rotation is weak, $\mathcal{R}_\alpha > \mathcal{R}_\Omega$ and the $\mathcal{R}_\Omega$ term could be dropped and the resulting system is then called an α^2-dynamo. When all three field-generating terms are retained we have an $\alpha^2\Omega$-dynamo.

6.4.2 Solutions of the Local Dynamo Equations

Equations (6.36) and (6.37), with suitable boundary conditions, describe an eigenvalue problem with solutions of the form $B_r(z, t; k) = f(z; k) \exp(\gamma_k t)$ and $B_\phi(z, t; k) = g(z; k) \exp(\gamma_k t)$, where the parameter k identifies each eigenvalue/eigenfunction pair. The exponential growth rates are the eigenvalues; if γ is complex the imaginary part describes oscillations against the background exponential growth $(\mathrm{Re}(\gamma) > 0)$ or decay $(\mathrm{Re}(\gamma) < 0)$. However, despite the many simplifications we have made to derive the basic dynamo equations, it is still not straightforward to find analytic forms for the solutions. In order to illustrate two important ideas we shall make further assumptions before summarising the results obtained by numerical methods. For simplicity, in the rest of the chapter we shall only consider the $\alpha\Omega$-dynamo, dropping the $\mathcal{R}_\alpha$ term in Eq. (6.37), so our dynamo model is described by

$$\frac{\partial B_r}{\partial t} = -\mathcal{R}_\alpha \frac{\partial}{\partial z}(\alpha B_\phi) + \frac{\partial^2 B_r}{\partial z^2}, \tag{6.39}$$

$$\frac{\partial B_\phi}{\partial t} = \mathcal{R}_\Omega B_r\, r \frac{\mathrm{d}\Omega}{\mathrm{d}r} + \frac{\partial^2 B_\phi}{\partial z^2}. \tag{6.40}$$

6.4.2.1 The Dynamo Number

Let us assume that $\alpha(z)$ is a constant, which does not violate the constraint that α is an odd function of z so long as we only consider one half of the disc. This allows us to find a simple analytic form of the eigenvalue γ. We can treat $B_r(z, t)$ and $B_\phi(z, t)$ as normal mode perturbations to an initial state of no magnetic field,

$$B_r(z, t) = \sum_{k=-\infty}^{\infty} \hat{B}_r \exp(ikz + \gamma t), \quad B_\phi(z, t) = \sum_{k=-\infty}^{\infty} \hat{B}_\phi \exp(ikz + \gamma t), \tag{6.41}$$

where i is the imaginary unit, and we shall assume for simplicity that $\alpha = 1$ and $r\mathrm{d}\Omega/\mathrm{d}r = -1$, so they take the reference values (for a flat rotation curve $\mathrm{d}\Omega/\mathrm{d}r < 0$) used when we non-dimensionalised the equations. What condition is necessary to make the real part of the growth rate positive, $\mathrm{Re}(\gamma) > 0$?

Substituting Eq. (6.41) into Eqs. (6.39) and (6.40) gives

$$\begin{bmatrix} \gamma + k^2 & ik\mathcal{R}_\alpha \\ \mathcal{R}_\Omega & \gamma + k^2 \end{bmatrix} \begin{bmatrix} \hat{B}_r \\ \hat{B}_\phi \end{bmatrix} = 0,$$

which has non-trivial solutions

$$\gamma = -k^2 \pm \sqrt{-ik\mathcal{R}_\alpha\mathcal{R}_\Omega} = -k^2 \pm \frac{1}{\sqrt{2}}(1+i)\sqrt{k\mathcal{R}_\alpha\mathcal{R}_\Omega}. \tag{6.42}$$

When $k = 0$ we have $\gamma = 0$. For $\gamma > 0$ the inequality $\mathcal{R}_\alpha\mathcal{R}_\Omega > 2|k|^3$ must be satisfied. Since the normal mode perturbations contain all k's the least stable mode is $|k| = 1$ and so if $\mathcal{R}_\alpha\mathcal{R}_\Omega > 2$ at least one of the modes grows exponentially in time and we have a dynamo. The behaviour of the system is governed by the dimensionless number

$$\mathcal{D} = \mathcal{R}_\alpha\mathcal{R}_\Omega, \tag{6.43}$$

which is called the dynamo number. The dynamo works when $|\mathcal{D}|$ exceeds a threshold, or critical, value $|\mathcal{D}| > |\mathcal{D}_c|$ with $|\mathcal{D}_c| = 2$ for this crude model.

Using Eqs. (6.35) and (6.19) we can estimate the magnitude of a typical dynamo number in a galactic disc as

$$\mathcal{D} \sim \frac{l^2\Omega_0}{\eta_T}\frac{\Omega_0 h^2}{\eta_T} \sim 2\left(\frac{l}{100\,\mathrm{pc}}\right)^2\left(\frac{h}{500\,\mathrm{pc}}\right)^2\left(\frac{v}{100\,\mathrm{km\,s^{-1}}}\right)^2\left(\frac{r}{10\,\mathrm{kpc}}\right)^{-2}, \tag{6.44}$$

which, for example, gives $\mathcal{D} \approx 10$ for the solar position in the Milky Way. In Sect. 6.4.2.4 we shall see that realistic critical dynamo numbers are also of the order of 10, depending on the exact form of $\alpha(z)$, and so we expect that $\mathcal{D}$ in galactic discs is generally close to the critical value.

If the alpha-effect and/or the shear are functions of the radius the behaviour of the solutions depends on the *local dynamo number* at the radius r, $\mathcal{D}_r = \mathcal{R}_\alpha\mathcal{R}_\Omega \alpha\, rd\Omega/dr$, where α and Ω are the local, non-dimensional, quantities. Note that if the galactic disc is flared then the dependence of $\mathcal{R}_\alpha$ and $\mathcal{R}_\Omega$ on $h(r)$ will also enter the local dynamo number.

6.4.2.2 Symmetry of the Solutions

The solutions $B_r(z)$ and $B_\phi(z)$ to Eqs. (6.36) and (6.37) can be grouped into two classes: those with even symmetry in z, $B_r(z) = B_r(-z)$ and $B_\phi(z) = B_\phi(-z)$ (and also $B_z(z) = -B_z(-z)$), and those with odd symmetry, $B_r(z) = -B_r(-z)$ and $B_\phi(z) = -B_\phi(-z)$ (and also $B_z(z) = B_z(-z)$). These classes are often referred to as quadrupole and dipole solutions, respectively. The symmetry class that dominates will be the one which contains the mode k, which has the highest growth rate $\mathrm{Re}(\gamma_k)$. Since the symmetry of the mean field is observable, at least in principle, this provides an important criterion to compare the theory and observations.

To gain some insight into how the growth rates of solutions with the different symmetries differ, we can solve Eqs. (6.39) and (6.40) with the generating terms removed (these are the terms with the coefficients $\mathcal{R}_\alpha$ and $\mathcal{R}_\Omega$). These solutions are called the free-decay modes, since the resulting two independent diffusion equations only have solutions which decay exponentially with time.

First we need to specify some boundary conditions for B_r and B_ϕ. The simplest and most commonly used boundary conditions are those that arise when the disc has sharp boundaries. Physically this can occur when, for example, there is a much higher diffusivity η_T outside in a surrounding halo. These are called the vacuum boundary conditions,

$$B_r(-h) = B_r(h) = B_\phi(-h) = B_\phi(h) = 0. \tag{6.45}$$

For these boundary conditions there are even-symmetry solutions to Eqs. (6.39) and (6.40) of the form

$$B_r = \hat{B}_r \cos\left[\left(k + \frac{1}{2}\right)\pi z\right]\exp\gamma_k t,$$

and an analogous equation for B_ϕ, which have growth rates

$$\gamma_k = -\left(k + \frac{1}{2}\right)^2 \pi^2, \quad k = 0, 1, 2, \ldots, \tag{6.46}$$

so the mode $k = 0$ decays slowest, at the rate $\gamma_0 = -\pi^2/4$. The odd-symmetry modes are of the form

$$B_r = \hat{B}_r \sin[k\pi z]\exp\gamma_k t,$$

with an analogous equation for B_ϕ. Note that here $k = 0$ gives the trivial solution $B = 0$, so the growth rates are

$$\gamma_k = -k^2\pi^2, \quad k = 1, 2, 3, \ldots, \tag{6.47}$$

and so the slowest decaying mode is $k = 1$ at the rate $\gamma_1 = -\pi^2$.

This means that since the slowest-decaying even-symmetry mode decreases four times less quickly than the slowest-decaying odd-symmetry mode, even-symmetry B_r and B_ϕ will dominate the free-decay solutions. The relative persistence of the even-symmetry modes may also apply to the solutions of the full equations, i.e. with the field-generation terms included, and so we might expect that even symmetry is preferred in this case too. This is confirmed below where we show numerical solutions of Eqs. (6.39) and (6.40). There is some support from observations that even vertical symmetry is preferred (see Sect. 5.2.3).

6.4.2.3   Magnetic Field Pitch Angles

The mean-field dynamo must generate both B_r and B_ϕ and so the regular magnetic field will have a non-zero pitch angle $p_B \neq 0$, where the pitch angle is given by Eq. (6.1) and is easy to measure in galaxies that are not edge-on, if the inclination of the disc is known. Table 5.3 shows p_B for a few nearby galaxies and $|p_B| > 0°$ is one of the strongest indications that a mechanism like the mean-field dynamo is responsible for generating the observed regular magnetic fields. There is thus a close connection between galactic dynamo theory and observations that is unique among all astrophysical dynamos; one cannot directly observe the dynamo active region in the Sun, stars and planets and so the link between dynamo theories for these objects and observable quantities is harder to establish.

To see why the galactic dynamo must produce both B_r and B_ϕ, consider the effect of setting $\mathcal{R}_\alpha = 0$ in Eq. (6.39). Then there are only exponentially decaying solutions, so that $B_r \to 0$ rapidly. This means that once $B_r \sim 0$, no matter how large $\mathcal{R}_\Omega$ is in Eq. (6.37), $B_\phi \to 0$ as there is no longer any B_r to be amplified by the Ω-effect. A similar result follows if we set $\mathcal{R}_\Omega = 0$ in Eq. (6.40).

We can estimate p_B from Eqs. (6.36) and (6.37) by setting $\alpha = 1$ and $r\,d\Omega/dr = -1$, approximating the z-derivatives as $\partial B/\partial z \sim B$ (remember z is scaled in units of h), $\partial^2 B/\partial z^2 \sim -B$ (we expect that there is a maximum in B_r and B_ϕ at the mid-plane, which is where we want to estimate p_B), and setting $B_{r,\phi}(t) \propto \exp \gamma t$. This gives the coupled pair of equations

$$(\gamma + 1)B_r + \mathcal{R}_\alpha B_\phi = 0,$$

$$(\gamma + 1)B_\phi + \mathcal{R}_\Omega B_r = 0,$$

which has non-trivial solutions when its determinant is zero and so $(\gamma + 1) = \sqrt{\mathcal{R}_\alpha \mathcal{R}_\Omega}$ (for a working dynamo we also require $\gamma > 0$, hence only the positive square root is retained). Then re-arranging to solve for the pitch angle gives

$$\tan p_B = \frac{B_r}{B_\phi} \sim -\sqrt{\frac{\mathcal{R}_\alpha}{\mathcal{R}_\Omega}}. \tag{6.48}$$

Note that both $\mathcal{R}_\alpha$ and $\mathcal{R}_\Omega$ depend on the disc scale-height h (Eq. 6.35), so that variations in h due to disc flaring, as is observed beyond the solar position in the Milky Way, will lead to a systematic variation of p_B with radius. The expected decrease in p_B with increasing radius has indeed been observed in several galaxies (Table 5.3).

Figure 6.4 shows observed and estimated pitch angles, where the estimates were obtained using Eqs. (6.19), (6.21), (6.35) and (6.48) and observed rotation curves, for two nearby galaxies. Even though the estimated pitch angles are based on a crude estimate, they are of the correct magnitude and even reproduce reasonably well the radial variation of the observed pitch angles.

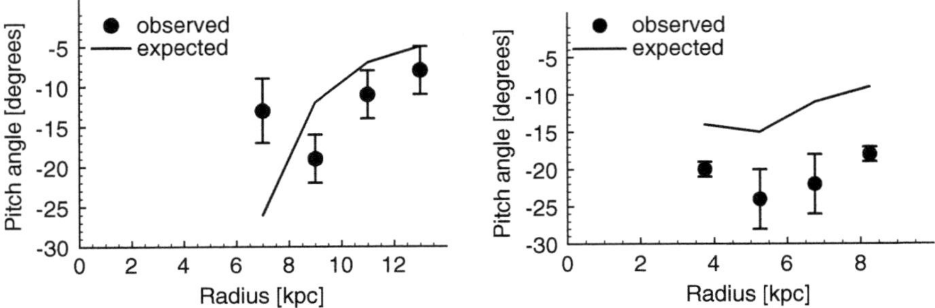

Fig. 6.4 Pitch angles of the axisymmetric ($m = 0$) regular magnetic field component derived from observations (points with error bars) and estimated using Eq. (6.48) (*lines*): *Left*: M31, *Right*: M51. To obtain the estimated p_B observationally derived rotation curves were used for each galaxy along with galaxy specific estimates of the disc scale-height and its radial variation. $l = 100\,\mathrm{pc}$ was assumed for the correlation length of the turbulence (Reproduced from Fletcher (2010))

6.4.2.4 Numerical Solutions

Equations (6.39) and (6.40) can be easily solved numerically using standard discretisation and time-stepping methods: 2nd-order approximations to the spatial derivatives and a low order Runge-Kutta scheme are perfectly adequate. This allows the solutions (the eigenfunctions) to the complete problem to be obtained. Figure 6.5 shows how the growth rate γ depends on the dynamo number $\mathcal{D}$. Here $\alpha(z) = \sin \pi z$ has been chosen, which produces the required vertical symmetry for α and $r\,d\Omega/dr = -1$ has been used for a flat rotation curve. For this form of $\alpha(z)$ the numerically obtained critical dynamo number is $\mathcal{D}_c = -8$ (note that there is a weak dependence of $\mathcal{D}_c$ on the form of $\alpha(z)$: for example, for $\alpha(z) = z$, $\mathcal{D}_c = -13$).

For a mildly supercritical dynamo number $\mathcal{D} \approx -10$ the dimensionless growth rate of $1/\gamma \approx 1$ represents a dimensional growth time $1/\gamma \approx 750\,\mathrm{Myr}$, if we take $h = 500\,\mathrm{pc}$. Thus for a galactic lifetime of $10\,\mathrm{Gyr}$ the mildly supercritical $\alpha\Omega$ dynamo undergoes about 13 e-foldings, increasing the field strength by a factor of order 10^5. In order to achieve current regular magnetic field strengths of $1\,\mu\mathrm{G}$ therefore requires an initial seed field of around 10^{-11} G. Many different electromagnetic effects in the early universe have been proposed as possible origins for the weak initial magnetic field upon which the dynamo acts; these can typically produce fields of order 10^{-17} G (see e.g. Widrow, 2002). The "missing" six orders of magnitude may be accounted for by additional amplification due to, for example: gas compression during the formation of the galactic disc; dynamo action in the early generations of stars or in AGN and their subsequent ejection; the fluctuation dynamo (see Sect. 6.5).

Figure 6.6 shows solutions obtained for the case of $\alpha(z) = \sin \pi z$ using two different dynamo numbers, $\mathcal{D} = -10$ and $\mathcal{D} = -100$. The signs of B_r and B_ϕ are not determined by the dynamo equations, but by the initial (weak) seed magnetic field. In Fig. 6.6 the seed field was random.

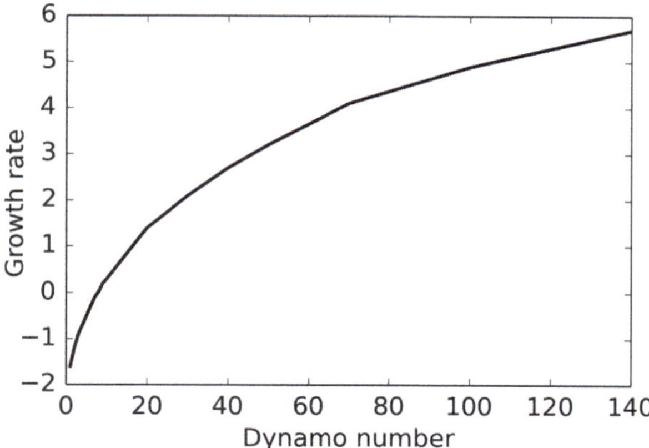

Fig. 6.5 Growth rate γ against the magnitude of the dynamo number $|\mathcal{D}|$ from numerical solutions to Eqs. (6.36) and (6.37) for $\alpha(z) = \sin \pi z$

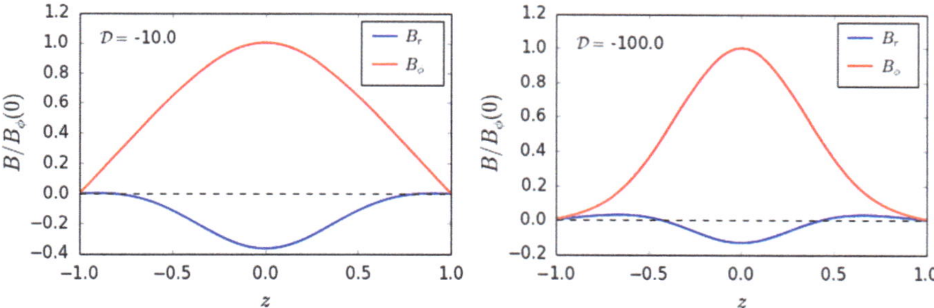

Fig. 6.6 Numerical solutions for the radial and azimuthal components of the regular magnetic field, as functions of z, for the case of $\alpha = \sin(\pi z)$, $\mathcal{R}_\alpha = 1$ and: (*left*) $\mathcal{R}_\Omega = -10$, (*right*) $\mathcal{R}_\Omega = -100$. Field strengths are normalised to the mid-plane value of B_ϕ

As expected $|B_\phi| > |B_r|$, since shearing of B_r by the differential rotation to generate B_ϕ is a more effective mechanism than the α-effect is for creating B_r from B_ϕ. As a consequence p_B will be small; in the case of $\mathcal{D} = -10$, $p_\mathrm{B} \approx -20°$, but for $\mathcal{D} = -100$, which is achieved by increasing $\mathcal{R}_\Omega$ the resulting field is more tightly wound, with $p_\mathrm{B} \approx -10°$.

The vertical symmetry of the dominant eigenfunction is even, i.e. $B_r(z) = B_r(-z)$ and $B_\phi(z) = B_\phi(-z)$, as anticipated from our analytic derivation of the growth rates of the decaying solutions. The exact shape of the eigenfunction depends on $\mathcal{D}$. Higher $\mathcal{D}$ results in the magnetic field being more concentrated towards the disc mid-plane and B_r changing sign near to the disc boundaries. The latter could produce an observable effect on p_B, the magnitude of which will depend on the vertical distribution of the synchrotron emissivity.

6.4.3 The Radial and Azimuthal Distribution of the Magnetic Field

So far we have only considered local solutions to the axisymmetric mean-field dynamo equations at a fixed radius. Here we shall derive an equation which describes the radial distribution of the mean-field dynamo generated magnetic field. This is the function $Q(r)$ in the separation of variables we introduced in Eq. (6.31).

The starting point is the dimensional Eq. (6.28), which describes the evolution of $B_r(r, z, t)$ in an axisymmetric disc. We introduce dimensionless variables, using the same scalings as in Sect. 6.4.1, but now retaining the term which originated in the radial component of the Laplacian. This gives

$$\frac{\partial B_r}{\partial t} = -\mathcal{R}_\alpha \frac{\partial}{\partial z}(\alpha B_\phi) + \frac{\partial^2 B_r}{\partial z^2} + \lambda^2 \frac{\partial}{\partial r}\left[\frac{1}{r}\frac{\partial}{\partial r}(r B_r)\right],$$

where all variables are dimensionless and $\lambda = h/R_0$. From Eq. (6.39), with $B_r(z, t) = \tilde{B}_r(z) \exp(\gamma t)$, we already have

$$\gamma \tilde{B}_r(z) = -\mathcal{R}_\alpha \frac{\partial}{\partial z}(\alpha \tilde{B}_\phi) + \frac{\partial^2 \tilde{B}_r}{\partial z^2},$$

where $\tilde{B}_r(z)$ and $\tilde{B}_\phi(z)$ are the local solutions, with local growth rate $\gamma(r)$, at each fixed position r. Now set

$$B_r(r, z, t) = Q(r)\tilde{B}_r(z) \exp(\Gamma t)$$

to obtain

$$[\gamma(r) - \Gamma]Q + \lambda^2 \frac{\partial}{\partial r}\left[\frac{1}{r}\frac{\partial}{\partial r}(rQ)\right] = 0. \tag{6.49}$$

A suitable outer boundary condition is $Q(R_0) = 0$, i.e. that the magnetic field is confined to the galactic disc. For the inner boundary, $Q(0) = 0$ is also usually chosen.

Although the galactic disc has been specified to be axisymmetric, we can also construct a dynamo model that allows for non-axisymmetric magnetic field components B_r and B_ϕ by writing

$$B_r(r, \phi, z, t) = Q(r)\tilde{B}_r(z) \exp(\Gamma_m t + im\phi),$$

and a similar equation for B_ϕ, where the m defines an azimuthal Fourier mode. An axisymmetric magnetic field has $m = 0$, a bisymmetric field has $m = 1$

(i.e. one maximum and one minimum in the field amplitude around a circle of radius) and so on. Clearly, if the mean-field dynamo is responsible for generating the observed regular magnetic fields discussed in Chap. 5 then those modes favoured by the dynamo, that is those with the highest growth rates and wide domains in the disc, should be the most commonly observed.

To summarise. The procedure to find the distribution of the dynamo generated magnetic field across a galactic disc can be broken down into a two step process. First, Eqs. (6.36) and (6.37) are solved at each radius, giving the local growth rate $\gamma(r)$ and the local field components $\tilde{B}_r(z)$ and $\tilde{B}_\phi(z)$. Then Eq. (6.49) is solved for $Q(r)$, which describes how diffusion of the magnetic field in the radial direction affects the solutions. Constructing the solutions to Eq. (6.49), especially for non-axisymmetric fields, requires some elaborate and lengthy mathematics, so we shall only report here on the two main results obtained by Baryshnikova et al. (1987).

Figure 6.7 (left panel) shows the growth rates Γ_m for three different azimuthal symmetries of the regular magnetic field. The most important result is that the axisymmetric mode, $m = 0$, has a higher exponential growth rate than modes with higher m values. It is also noticeable that as the annular velocity increases, thus producing a higher $\mathcal{R}_\Omega$, the growth rates of the bisymmetric ($m = 1$) and doubly peaked ($m = 2$) modes decrease significantly. Thus we should expect that for typical disc galaxies the mean-field dynamo will preferentially generate B_r and B_ϕ that are symmetric under rotation.

The right panel of Fig. 6.7 shows that the $m = 0$ mode extends over a greater radial range in the disc than the higher modes, which are localised in distinctly

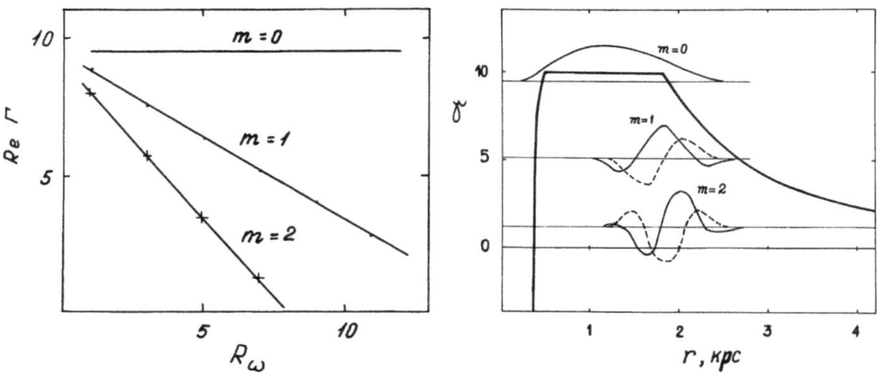

Fig. 6.7 Solutions of Eq. (6.49) for magnetic fields with azimuthal modes $m = 0$, $m = 1$ and $m = 2$, $\alpha = 1$ for $z > 0$, $\alpha = -1$ for $z < 0$, $\mathcal{R}_\alpha = 1$ and $\lambda = 0.04$. *Left*: growth rates Γ of the different modes m as a function of $\mathcal{R}_\Omega$. *Right*: growth rates as a function of galactic radius for a model of the galaxy M 51. *Thick solid line*: local growth rate $\gamma(r)$. *Thin solid lines*: growth rates of the different azimuthal modes Γ_m. *Solid* and *dashed curves*: the real and imaginary parts, respectively, of the eigenfunction Q_m, these show the radial range over which the mode is localised (From Baryshnikova et al. (1987))

smaller radial ranges. So, again, the mean-field theory predicts that axisymmetric regular magnetic fields should be preferentially found in the observations as they are excited over a wider radial range in a typical galactic disc. This is indeed what is found in the observations listed in Table 5.4 and this provides strong evidence that a mean-field dynamo, of the type derived in Sect. 6.4, is responsible for producing the observed regular magnetic fields.

6.4.4 Non-linearity and Dynamo Saturation

A kinematic dynamo, where the velocity is not modified by the magnetic field, has no mechanism to switch itself off. Thus, if the dynamo number exceeds the critical value the magnetic field will continuously grow at an exponential rate. This is clearly unphysical, since at some point the Lorentz force must become strong enough to alter the properties of the turbulent velocity field that the kinematic dynamo model assumes to be constant.

The simplest and most common non-linearity is called α-quenching, which makes the α-effect time dependent by defining it as

$$\alpha = \frac{\alpha_0}{1 + B^2/B_{eq}^2}, \tag{6.50}$$

where $B_{eq}^2 = 4\pi\rho v'^2$, with ρ the gas density, is called the equipartition field strength, which replaces the arbitrary B_0 as the measure of magnetic field in Sect. 6.4.1. This quenching is motivated by both theoretical considerations, that the magnetic field will only affect $\vec{v}'$ when its energy density approaches the turbulent energy density, and observations, which show that an approximate equipartition of energy density between turbulence and the magnetic field holds. In this form the α-effect becomes weaker as $B \rightarrow B_{eq}$, thus the dynamo becomes less efficient and reaches a steady strength of $\sqrt{B_r^2 + B_\phi^2} \approx B_{eq}$.

Since the diffusivity η_T of Eq. (6.22) is also due to turbulent effects it should also be sensitive to a growing field strength. Thus sometimes η-quenching, of a form similar to Eq. (6.50) is also introduced.

One of the most controversial aspects of mean-field dynamo theory is how strong this α-quenching might be in reality. We shall see below, in Sect. 6.5, that small-scale magnetic fields can be very efficiently amplified by a turbulent flow, in a process that is independent of the mean-field dynamo. So it can be argued that saturation of the mean-field dynamo, due to a rapid increase in $\vec{B}'$ suppressing the helicity of $\vec{v}'$ that is required for the α-effect, will occur much earlier than Eq. (6.50) suggests. Some numerical simulations suggest that the magnetic Reynolds number, raised to some power, should be a coefficient of the term $(B/B_{eq})^2$. Since $\mathcal{R}m \gg 1$, this would lead to a much stronger quenching of α and so the dynamo would saturate with $B \ll B_{eq}$. This suggestion of a "catastrophic" α-quenching has led to a search for different saturation mechanisms that take into account more carefully other important aspects of galactic astrophysics.

α-quenching is obviously a crude non-linearity that is only phenomenologically connected to the underlying physics. An alternative approach is to *derive* an equation describing the evolution of α with a dependence on the properties of the turbulence and the regular magnetic field. For example, Shukurov et al. (2006) argue that since magnetic helicity (defined as $(\nabla \times \vec{B}) \cdot \vec{B}$) is conserved in a closed system and the dynamo produces regular magnetic fields with a common sign of helicity, there must also be a build-up of small-scale field with the opposite sign of magnetic helicity. This increase in small-scale magnetic helicity will affect the helicity of the turbulent velocity field that is responsible for the α-effect, but this effect can be alleviated if a galactic wind or fountain transports the small-scale magnetic helicity out of the disc. These arguments are used to derive an equation for $\partial \alpha / \partial t$ that is solved simultaneously with the evolution equations for B_r and B_ϕ.

6.5   Small-Scale Magnetic Fields and the Fluctuation Dynamo

Observations show that as well as a regular magnetic field, ordered on scales of 1 kpc or more, the interstellar medium also contains random fields with smaller correlation scales. There are at least two possible mechanisms which can produce the random magnetic fields: tangling up of the regular field generated by the mean-field dynamo and direct generation of random fields by the fluctuation dynamo.

The fluctuation dynamo mechanism is independent of the mean-field dynamo and applies to all flows which are turbulent: random stretching of magnetic field lines by the turbulent velocity field leads to a rapid local growth of small-scale magnetic energy. A mathematical analysis of the fluctuation dynamo is lengthy and difficult so we will only briefly describe the two most important features of the process, which can be useful in interpreting observations.[2]

First, the exponential growth time, $1/\gamma$, of the small-scale magnetic field is of the order of the "turnover" time of the turbulent eddies: $1/\gamma \sim l/v' \sim 100\,\mathrm{pc}/10\,\mathrm{km\,s^{-1}} = 10\,\mathrm{Myr}$. Thus the growth rate is 10–100 faster than that of the mean-field dynamo. How this rapid amplification of small-scale magnetic field affects the operation of the $\alpha\Omega$-dynamo, in particular the possibility that the α-effect will be quenched before the mean magnetic field has had a chance to be significantly amplified, is a topic of much current debate. Brandenburg and Subramanian (2005) is a good starting point for the interested reader.

---

[2]The most accessible technical description of the fluctuation dynamo is probably that in Zel'dovich et al. (1990). Brandenburg and Subramanian (2005) discuss the theory as well as the results of numerical simulations.

Another significant property of the fluctuation dynamo is the character of the small-scale magnetic fields that it produces. It can be shown that the growth rates for higher moments of the magnetic field are faster than for lower moments: for example the growth rate of B'^2 is about three times that of B' (Zel'dovich et al., 1990). This means that the magnetic field is not uniformly spread throughout the volume occupied by a turbulent cell, but is concentrated in randomly oriented, thin magnetic flux "ropes", whose characteristic thickness is $d \sim l/\sqrt{\mathcal{R}m}$. This *intermittent*, rather than smooth, distribution of the fluctuation dynamo generated small-scale fields is different from the fields produced by tangling up the mean field, which will be uniformly distributed in space.

6.6 Further Reading

The book by Choudhuri (1998) is an excellent introduction to fluid dynamics, including turbulence and magnetohydrodynamics, in an astrophysical context. At a more advanced level Kulsrud (2005) provides a comprehensive overview of astrophysical plasmas. For a clear and thorough discussion of hydrodynamic turbulence, Tennekes and Lumley (1972) is highly recommended. The review papers Beck et al. (1996) and Widrow (2002) set out the connection between observations and dynamo theory and discuss many questions which are still to be answered. The standard reference on galactic dynamo theory for many years has been Ruzmaikin et al. (1988): a new edition is forthcoming. Kulsrud and Zweibel (2008) take a critical look at the standard $\alpha\Omega$-dynamo theory. Review papers based on lectures on galactic dynamos (Shukurov, 2007) and dynamo theory (Jones, 2008) are good places to delve into more detail, as is the comprehensive review on astrophysical dynamos by Brandenburg and Subramanian (2005), which covers the current state of the field in great depth.

References

Baryshnikova, Y., Ruzmaikin, A.A., Sokoloff, D.D., Shukurov, A.M., 1987, Astron. Astroph. **177**, 27

Beck, R., Brandenburg, A., Moss, D., Shukurov, A., Sokoloff, D., 1996, ARA&A **34**, 155

Brandenburg, A., Subramanian, K., 2005, Physics Reports, **417**, 1

Choudhuri, A.R., 1998, *The Physics of Fluids and Plasmas: an Introduction for Astrophysicists*, (Cambridge: CUP)

Fletcher, A., 2010, in *The Dynamic Interstellar Medium: A Celebration of the Canadian Galactic Plane Survey* eds. R. Kothes, T.L., Landecker, A.G., Willis, ASP Conference Series **438**, 197

Jones, C.A., 'Dynamo Theory' in Les Houches Volume 88 Dynamos 2008, eds. Ph. Cardin and L.F. Cugliandolo (Les Houches: Elsevier)

Kulsrud, R.M., 2005, *Plasma Physics for Astrophysics* (Princeton: Princeton University Press)

Kulsrud, R.M., Zweibel, E.G., 2008, Reports on Progress in Physics **71**, 046901

Parker, E.N., 1955, Astroph. J. **122**, 293

Parker, E.N., 2007, Conversations on Electric and Magnetic Fields in the Cosmos (Princeton: Princeton University Press)

Ruzmaikin, A.A., Shukurov, A.M., Sokoloff, D.D., 1988, *Magnetic Fields of Galaxies* (Dordrecht: Kluwer)

Shukurov, A., 2007, '*Introduction to galactic Dynamos*' in Mathematical Aspects of Natural Dynamos, eds. E. Dormy, A. Soward, (Francis: CRC Press/Taylor)

Shukurov, A., Sokoloff, D., Subramanian, K., Brandenburg, A., 2006, Astron. Astroph. **448**, L33

Steenbeck, M., Krause, F., Rädler, K.-H., 1966, Zeitschr. Naturforsch. Teil A **21**, 369

Tennekes, H., Lumley, J.L., 1972, *A First Course in Turbulence*, (Cambridge: MIT Press)

Widrow, L.M., 2002, Reviews of Modern Physics, **74**, 775

Zel'dovich, Ya.B., Ruzmaikin, A.A., Sokoloff, D.D., 1990, *The Almighty Chance* (Teaneck N.J., World Scientific)

Chapter 7
Active Galactic Nuclei

7.1 The AGN Zoo

So far, we have discussed the radio continuum (specifically: the synchrotron) emission of normal, i.e. star-forming galaxies, along with their magnetic field. As pointed out in Sect. 5.5, their synchrotron luminosity is closely associated with the star formation in their disks, this giving rise to the well-known radio-FIR correlation established for star-forming galaxies.

This tight correlation does not hold, however, for galaxies hosting active galactic nuclei (AGN). They still obey such a law, however, one with a lot more scatter and with the radio luminosity being about two orders of magnitude higher (at the same FIR luminosity). In Fig. 7.1, the monochromatic radio luminosity at 5 GHz is plotted versus the FIR luminosity for two samples of galaxies, one comprised by star-forming galaxies and the other one containing galaxies with central AGN. A bimodal distribution is immediately evident, forming two separate correlations. In the lower part of the diagramme, we retrieve the 'normal' radio-FIR correlation as presented in Sect. 5.5. It exhibits less scatter than the other correlation visible in the upper part of the graph, with the latter representing AGN-dominated galaxies that have excess radio power at the same FIR luminosity. This implies that the FIR radiation is not emitted by the AGN, and that the powering of the AGN is by and large proportional to the FIR luminosity.

Another interesting finding resulting from this diagramme is that the 'normal' radio-FIR correlation also contains so-called radio-quiet quasars (see Sects. 7.5 and 7.2), which subtend only three orders of magnitude in luminosity, as well as the bulk of Seyfert galaxies (Sects. 7.6 and 7.2), which spread over a large luminosity range. The inference is that both radio-quiet quasars as well as Seyfert galaxies have normal spiral hosts in which star formation is at work, while radio-loud quasars and radio galaxies have elliptical host galaxies, with comparatively low FIR luminosities, and the radio power essentially produced by the central AGN.

© Springer International Publishing Switzerland 2015

U. Klein, A. Fletcher, *Galactic and Intergalactic Magnetic Fields*,
Springer Praxis Books, DOI 10.1007/978-3-319-08942-3_7

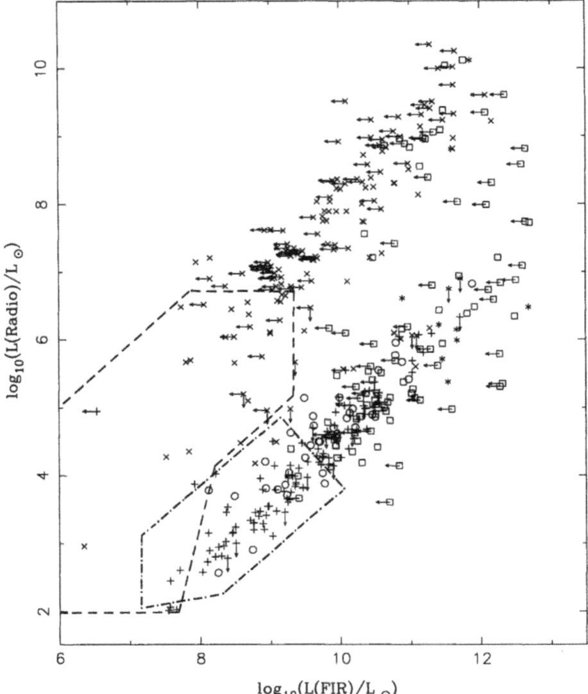

Fig. 7.1 Radio-FIR correlation of star-forming galaxies and galaxies hosting AGN (Sopp and Alexander, 1991)

The production of relativistic particles emitting their radio (synchrotron) emission in AGN is fundamentally different from that of star-forming galaxies. AGN show up in a variety of forms and have been identified as a new class of galactic systems in the second half of the last century. The first recognition that there are extreme phenomena in the centres of certain galaxies was made for the so-called Seyfert galaxies (Seyfert, 1943). With their two categories, viz. Sy 1 and Sy 2, they are templates for the standard AGN paradigm. Sy-1 galaxies exhibit broad emission lines, with $FWHM \sim 5000 \ldots 10\,000\,\mathrm{km\,s^{-1}}$, and a (partially) nonthermal, nonstellar continuum, recognised as synchrotron radiation. The central region emitting these broad spectral lines is therefore called 'broad-line region' (BLR). In contrast, Sy-2 galaxies show 'narrow' lines, $FWHM \sim 200 \ldots 400\,\mathrm{km\,s^{-1}}$, which is still broader than the corresponding linewidths of typical normal galaxies. The central regions of this type of Seyfert galaxies is called 'narrow-line region' (NLR). From the observed variability of the emission from Sy-1 galaxies one infers a linear size of the BLR $< 10^{16}$ cm (1/100th of a lightyear), while for NLR this is $\sim 10^2 \ldots 10^3$ times larger. The difference between the BLR in Sy 1 and the NLR Sy 2 is the viewing angle (see the so-called 'unified scheme' explained in Sect. 7.3).

After World War II, radio astronomy saw a fast development. Bolton and Stanley (1948) detected the radio source Cyg A (3C 405) with their 'Sea Interferometer'. Smith (1951) obtained a more accurate position at Cambridge, and Baade identified

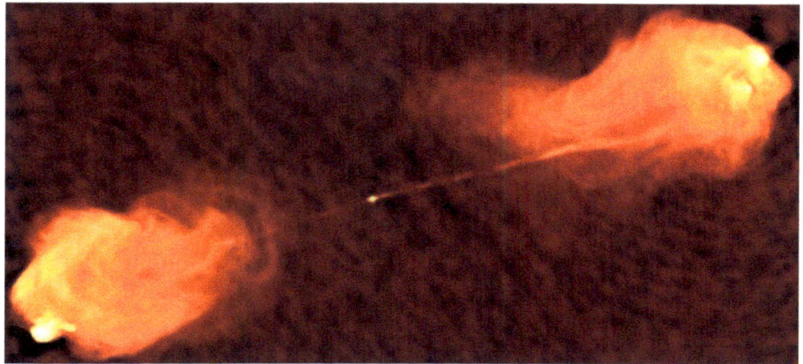

Fig. 7.2 Radio image of Cyg A at $\lambda = 6\,\text{cm}$ (Perley et al., 1984a)

the radio source with a distant ($D \approx 700$ million ly) optical galaxy (Baade and Minkowski, 1954). The inference therefore was that Cyg A (Fig. 7.2) is an extremely radio-luminous galaxy, producing $\sim 10^4$ times the radio luminosity of M 31. Schmidt (1963) identified the radio source 3C 273 with a 13th magnitude star-like object, having a redshift of $z = 0.158$ (obtained from the spectroscopy of the Balmer series). Matthews and Sandage (1963) had discovered a 16th magnitude star-like object at the position of the radio source 3C 48. However, they did not yet obtain a spectrum, hence there was no known redshift for this source at that time (1960s). The broad emission lines could not be identified, and the spectrum exhibited much more UV light than that of an ordinary main-sequence star.

7.2 Taxonomy of AGN

7.2.1 Quasars

Quasars are very luminous and compact centres of galaxies, outshining their host galaxies. They exhibit nearly featureless spectra, and are visible from radio wavelengths through X-ray energies. Their luminosities, dominated by synchrotron radiation, are $10^{45} \ldots 10^{49}\,\text{erg}\,\text{s}^{-1}$, exceeding up to a thousand times the total luminosity from a massive galaxy like the Milky Way. Perhaps the most stunning fact is that this luminosity is emitted from a volume not much larger than the solar system. To date, $\sim 10^4$ QSOs are known, of which about 10 % are strong radio emitters, so-called 'radio-loud' QSOs, while the majority exhibit similar properties, except for being 'radio-quiet'. Quasars constitute the most luminous objects in the universe (Sect. 7.5) and according to current knowledge are connected to a period of violent merging of massive galaxies in the course of structure formation of the universe, peaking in redshift range between $z \approx 2$ and $z \approx 0.4$.

7.2.2   *Radio Galaxies*

This kind of AGN phenomenon was first recognised in the form of its prototype Cyg A. The host galaxies are ordinary ellipticals in the optical regime. They are very strong radio emitters, with radio luminosities of $L_r \geq 3 \cdot 10^{41}$ erg s^{-1}. They mostly consist of a compact core, they show a jet or twin jets, which feed double lobes with a relativistic plasma. Sometimes they exhibit so-called 'hotspots' in their outer lobes. These radio galaxies show two morphologically different categories, which can also be subdivided by their radio luminosities (first recognised by Fanaroff and Riley, 1974):

- FR I radio sources have monochromatic radio luminosities of $L_{1.4\,GHz} \leq 10^{32}$ erg s^{-1} Hz^{-1}. They possess less collimated jets, which exhibit a loss of collimation beyond the optical peripheries of their host galaxies, with a strong flaring further out. They then become fainter and have steeper synchrotron spectra towards the outer edges of the lobes.
- FR II radio sources have monochromatic radio luminosities of $L_{1.4\,GHz} \geq 10^{32}$ erg s^{-1} Hz^{-1}. Their highly collimated jets are often one-sided, with pronounced hotspots at their termination points in the lobes. These lobes extend out of the host galaxies for many tens of kiloparsecs, one of the biggest such object known to date being the radio galaxy 3C 236, with a total extent of $\sim$4.5 Mpc ($H_0 = 71$ km s^{-1} Mpc^{-1})!

This difference is likely due to the jets in FR Is being subsonic, and supersonic in FR IIs. The difference in jet speed may be also connected with the different properties of the ISM in the host galaxies, i.e. its density in the first place.

7.2.3   *Seyfert Galaxies*

Seyfert galaxies are spiral galaxies hosting an AGN in their centre. Their luminosity is in the range $\sim 10^{43} \dots 10^{45}$ erg s^{-1}, with the brightest Seyfert galaxies being as luminous as the faintest QSOs. As already pointed out in Sect. 7.1, the main difference between the two types, Sy 1 and Sy 2, is that they show rather different line widths when looked at with optical spectroscopy:

- In a Sy 1 galaxy, the observer measures a large velocity dispersion, as one is looking directly at the immediate surroundings of the central supermassive black hole.
- In a Sy 2 galaxy, our view onto it has a different aspect angle, which causes the central region to be obscured by a dust torus.

Nearly 15 % of Seyferts have close companions, which is indicative of merging or tidal forces at work.

7.2.4 Blazars

Blazars emit polarised light with a featureless, nonthermal spectrum that can be traced over a large spectral range. They also exhibit very strong variability, therefore one subdivides them into two types. BL Lacs are named after its prototype BL Lacertae. These do not exhibit any emission lines at all. Optically violently variable sources (OVV) do show emission and absorption lines when subject to scrutiny. Blazars are most likely AGN in which we look almost directly into the jet coming towards us, which results in strong relativistic boosting, the so-called 'Doppler boosting'. This process enhances the unboosted flux density S_0 towards the observer to a value of

$$S_{obs} = S_0 \cdot D_+^{3+\alpha}, \tag{7.1}$$

where

$$D_+ = \frac{1}{\gamma \cdot (1 - \beta \cos \theta)},$$

is the Doppler factor,

$$\gamma = (1 - \beta^2)^{\frac{1}{2}}$$

is the Lorentz factor and

$$\beta = \frac{v}{c}.$$

The source emits synchrotron radiation with the usual power-law (in the rest frame):

$$S_{v_0} \propto v_0^{-\alpha}.$$

Correspondingly, the emission of the receding jet is diminished to

$$S_{obs} = S_0 \cdot D_-^{3+\alpha}, \tag{7.2}$$

where

$$D_- = \frac{1}{\gamma \cdot (1 + \beta \cos \theta)}.$$

This is the reason why radio galaxies generally show a weaker counter-jet, and why the counter-jet is mostly invisible in quasars. The extreme Doppler boosting in blazars is the reason why one does not see any line emission in their spectra. The line emission is simply swamped out by the vastly dominating and outshining, Doppler-boosted synchrotron radiation that we measure when looking into the jet. The variability is readily explained in terms of wiggeling jets, implying that the direction θ of the boosting w.r.t. the observer varies, hence producing a variable

Doppler factor. Using Eq. (7.1), it is readily seen that for β close to unity a slight variation of the angle θ leads to a strong variation of the observed flux density S_{obs}.

7.3 Unified Scheme

In this section, we summarise the properties of the AGN phenomenon, which is largely connected with the production and action of magnetic fields. A summary of the signs of nuclear activity is given below. It has to be pointed out that not all of these features are always present, but they have led to what is called the *unified scheme* (Urry and Padovani, 1995).

- There is a compact ($\leq 3\,$pc), luminous centre.
- It shows a spectrum with strong emission lines, and these are strongly Doppler-broadened.
- The central region emits strong nonthermal emission.
- Strong UV radiation from their compact central region is observed.
- Frequently, jets are formed that feed double radio sources.
- There may be variability of the emission from the central region over short timescales, which encompass the whole spectrum.

These properties can be explained in terms of the so-called 'unified scheme' (Fig. 7.3) of AGN, with the following ingredients:

- In the centre of these galaxies there is a supermassive black hole.
- The central black hole is surrounded by an accretion disk.
- Twin jets are ejected into two opposite directions.
- The BLR reflects the immediate surroundings of the black hole and the accretion disk.
- The NLR marks the gas motions further away from the compact nuclear source.
- There is a molecular and/or dust torus surrounding this central BLR which, depending on the viewing angle blocks the view onto the central region, hence showing only the NLR, while the view onto the BLR is unblocked in the other case.

The linear dimensions of the main AGN components are given in the tabular compilation below:

Component	Typical size
BH (R_S for $10^9\,M_\odot$)	$10^{-4}\,$pc
Accretion disk	0.01 pc
BLR	1 pc
Dust torus	10 pc
NLR	1 kpc
Host galaxy	10 kpc

Fig. 7.3 Sketch of the central region of a galaxy illustrating the unified scheme (From Urry and Padovani, 1995)

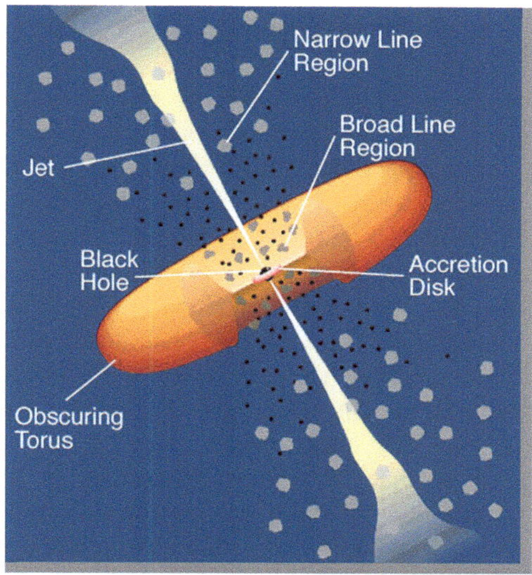

The brightness temperature

$$T_b = \frac{\lambda^2}{2\,k} \cdot \frac{S_\nu}{\Omega_S}$$

is an important tool to diagnose the radiation process of a source. As we have seen in Sect. 2.2.2, this quantity is defined via the Rayleigh-Jeans approximation of a radiation process. Nonthermal radiation processes are generally characterised by high brightness temperatures, which of course do not have any physical meaning, except that they tell something about the intensity of the source. For AGN the intensities can be extreme. However, the brightness temperature that a radio source can possess is limited by the fact that for very high number densities and energies of relativistic electrons, inverse-Compton losses become dramatic. The maximum brightness temperature is constrained by the ratio of inverse-Compton to synchrotron powers:

$$\frac{P_{IC}}{P_{syn}} = \frac{1}{2} \cdot \left(\frac{T_{max}}{10^{12}\,\mathrm{K}}\right)^5 \cdot \left(\frac{\nu_c}{\mathrm{MHz}}\right) \cdot \left[1 + \frac{1}{2} \cdot \left(\frac{T_{max}}{10^{12}\,\mathrm{K}}\right)^5 \cdot \left(\frac{\nu_c}{\mathrm{MHz}}\right)\right].$$

For temperatures $T > 10^{12}\,\mathrm{K}$, the inverse-Compton losses become catastrophic, as in this case $P_{IC}/P_{syn} \sim T_{max}^{10}$. This situation is referred to as the inverse-Compton catastrophe.

7.4 Radio Galaxies

Radio galaxies are probably the most well-defined type of AGN in terms of their radio structures, in spite of their bimodal distribution of morphologies and luminosities (FR I, FR II). They possess (almost) all of the ingredients mentioned so far, except that, or because of, their host galaxies are elliptical galaxies the ISM of which is a tenuous and hot one. If they are radio-loud, which holds true for all FR II radio galaxies, they can be observed out to very large distances, hence may also serve as cosmological probes. We will first discuss FR II sources, since they represent the radio-galaxy phenomenon as it had first been noted in the form of its prototype, Cyg A (3C 405), which shows the two jets ploughing through the ISM and IGM, finally piling up at the hotspots and, from there, depositing the relativistic plasma in the outer lobes.

Note that, in contrast to studies of normal galaxies, the radio continuum observations of radio galaxies do not suffer from the contamination by thermal free-free radiation, hence the interpretation is straightforward here.

7.4.1 FR II Sources

In the radio regime, the main features of FR II sources are as follows (see Fig. 7.2 and the sketch in Fig. 7.4):

- A central compact (pc-scale) radio source.
- Twin jets ejected in opposed directions (with different synchrotron brightness).
- The jet opening angle $\xi = dr_j/d\ell < 0.1$ rad, with typical values around 0.05 (r_j being the jet radius and ℓ the length).
- Hotspots marking the terminations of the jets.
- Double lobes, 'floating' back onto the host galaxies from the hotspots ('Plume' in Fig. 7.5).
- They are edge-brightened.
- They show radio luminosities at least as high as $L_{1.4\,\mathrm{GHz}} \geq 10^{32}$ erg s^{-1} Hz^{-1}.

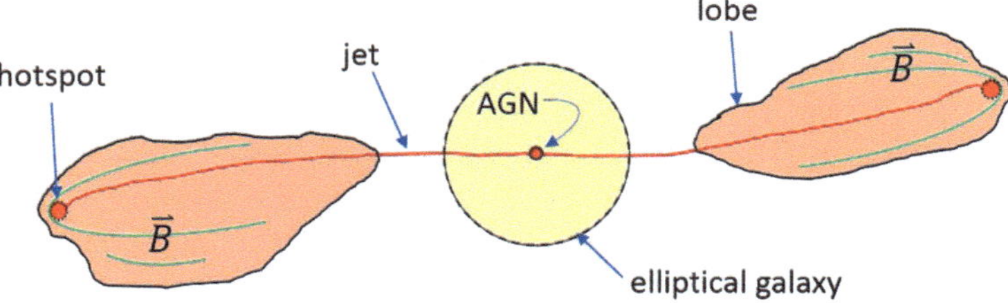

Fig. 7.4 Sketch of the main features of a classical FR II radio galaxy

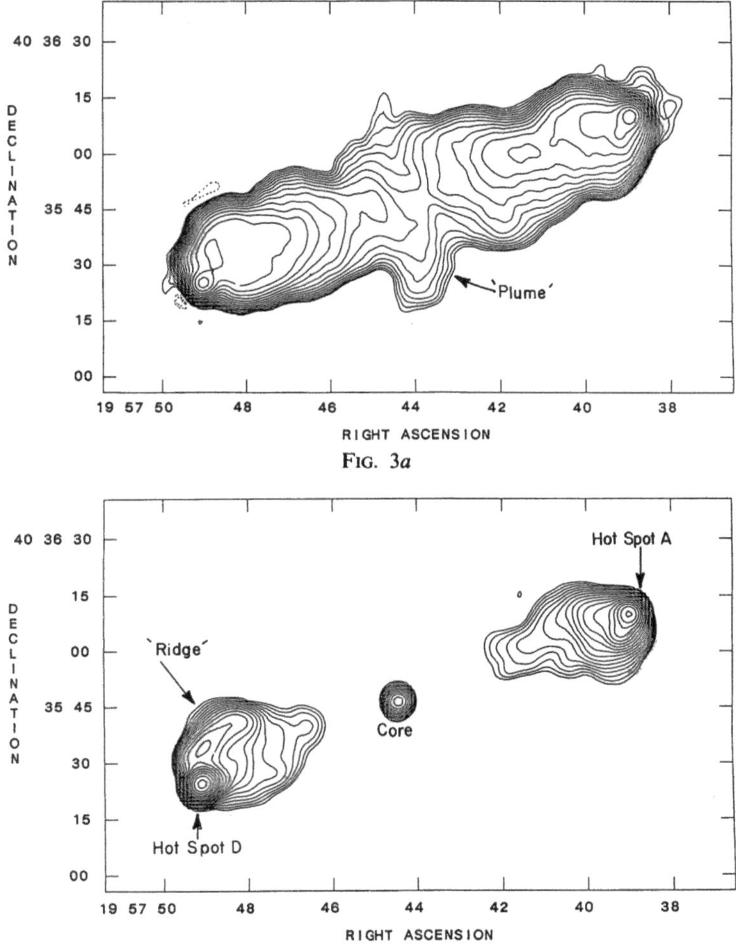

Fig. 7.5 Cyg A at 327 MHz (*top*) and at 15 GHz (*bottom*) (Carilli et al., 1991)

As said above, the most prominent representative of this species is the radio galaxy first discovered and identified as such, Cyg A. Since it is one of the radio galaxies that has been studied in enormous detail (Carilli and Harris, 1996), it is useful to discuss the properties of radio galaxies using Cyg A as a template. Since we are dealing with magnetic fields and hence with the emitted synchrotron radiation used to study radio galaxies, the appearance of Cyg A in the radio domain is first briefly outlined, as it has been studied in the radio continuum, even with well resolved linear polarisation on a wide range of scales, encompassing the jets and lobes. Cyg A was also the first radio galaxy for which the influence on the surrounding (hot) gaseous medium was discovered, as the jets, while propagating forward in the IGM, create cavities that are readily seen in the X-ray regime. At

X-ray energies, cavities driven into the hot thermal gas by the jets are evident (see Sect. 8.7). The most salient features of the template radio galaxy Cyg A are:

- Looking at Fig. 7.5, we see the basic morphology at this somewhat lower angular resolution, with a core and the outer lobes. Note, however, the different appearance at different frequencies: at the lower frequency, we see thick lobes, with aged particles streaming back onto the host galaxy, whereas at the higher frequency, we only see the high-energy particles, mainly from the jets and lobes.
- The magnetic field is mostly circumferential in the lobes, following their boundaries (Fig. 7.6). This is a general phenomenon seen in radio galaxies.
- There are indications of the magnetic field to be oriented parallel to the jet (like in most FR IIs). Within the lobes, the magnetic field is more tangled, with field-ordering on scales of $\sim 20 \ldots 30$ kpc.
- One frequently finds large rotation measures, ~ 4000 rad m^{-2} $<$ RM $<$ $+3000$ rad m^{-2}. These must be due to a foreground Faraday screen, i.e. the hot ICM gas and the intergalactic magnetic field.

The inferred intergalactic magnetic field has a strength of $B_{ICM} \sim 2 \ldots 10\,\mu$G (see Chap. 8 for more details on this topic). Below we list the strength of the magnetic field within Cyg A from the observed synchrotron intensity, assuming equipartition:

B_t (μG)	Location
100	Core
80	Jets
200	Hot-spots
70	Lobe's head
40	Lobe's tail

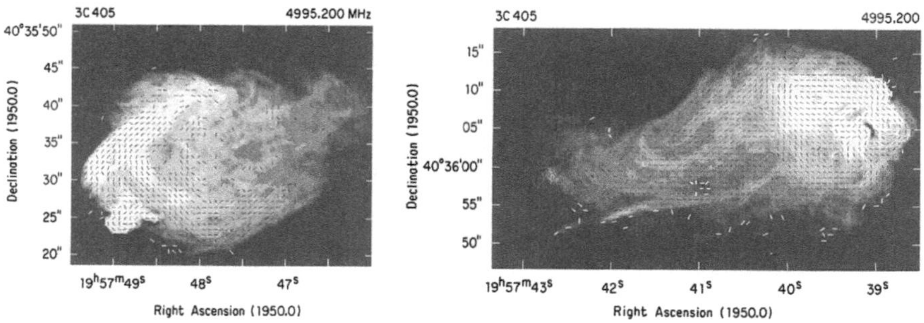

Fig. 7.6 The magnetic field in Cyg A as deduced from measurements of the linear polarisation (Dreher et al., 1987)

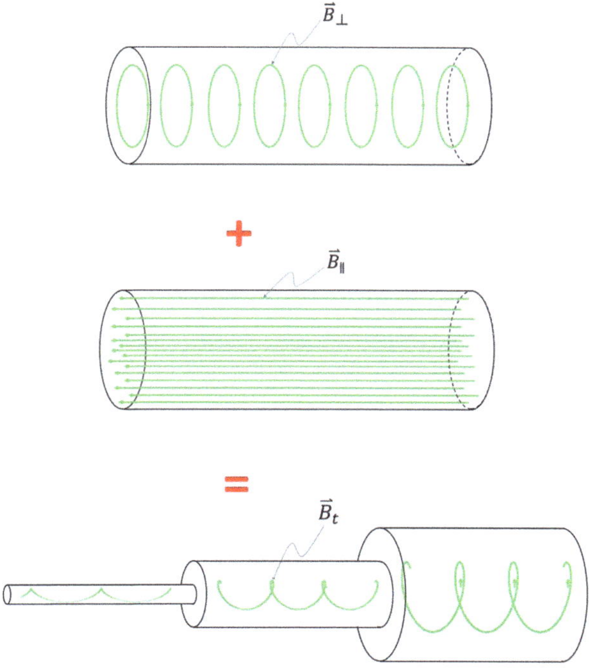

Fig. 7.7 Illustration explaining flux conservation in a flux rope made up by longitudinal and poloidal magnetic fields

The fact that in FR-II radio galaxies the magnetic field is parallel to the jet over most of its length can be explained in terms of the large bulk velocities of the jet material. This is perhaps best illustrated by the 'flux-rope' model (Fig. 7.7), which was 'invented' to explain the magnetic flux rope on the planet Venus. Flux conservation requires that

$$B_\parallel \propto r^{-2}, \qquad B_\perp \propto r^{-1}.$$

This is a strongly simplified picture, as it cannot account for the intricate physics, such as the entrainment of surrounding material from the ISM and/or ICM, or (oblique) shocks. In fact, for instance, the very prominent jet in the giant radio galaxy NGC 6251 exhibits a preferentially longitudinal magnetic-field orientation over the first (projected!) $\sim$40 kpc, while further out it is dominated by the perpendicular, or poloidal, component. A helical magnetic-field structure has meanwhile been inferred for several jets (3C 120, 3C 273, ...) from rotation measure analyses. A different morphology is seen in some pc-scale jets (e.g. the blazar 1055+018), which show two components of polarisation. An inner 'spine' with a transverse field is bracketed by a boundary layer having a longitudinal field. The interpretation here is that the inner spine is dominated by oblique shocks, while the jet's mantle interacts with the surrounding medium. This interaction can be

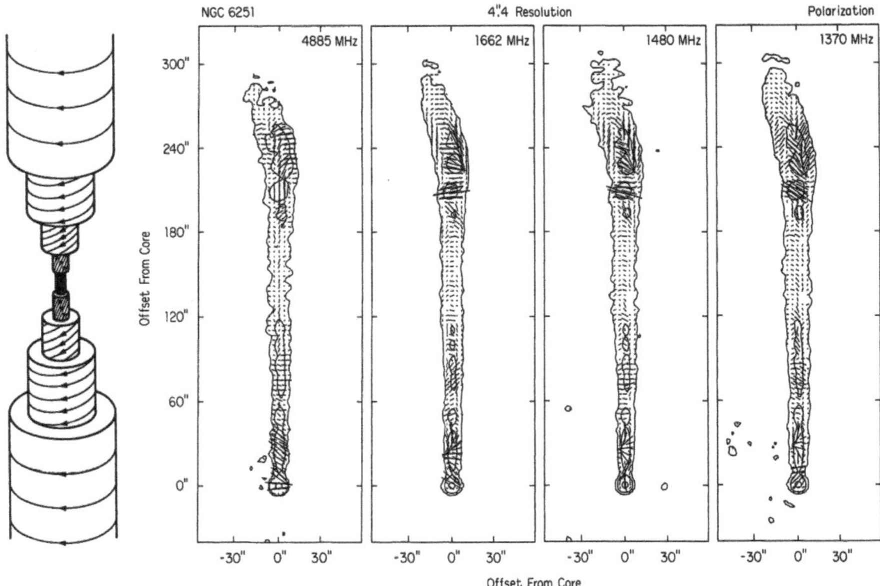

Fig. 7.8 Sketch of the magnetic field in a flux rope (Russell and Elphic, 1979), with a real magnetised jet (NGC 6251; Perley et al., 1984b) juxtaposed

twofold: (i) shear of the jet flow stretching the magnetic field along the jet, and (ii) the jet fluid can be compressed against its boundaries. In Fig. 7.8 the prominent jet in NGC 6251 is shown.

7.4.2   FR I Sources

FR I sources are quite different from FR IIs both, in terms of their luminosity and their morphology. The main features and properties of FR I radio galaxies are:

- They are edge-darkened, i.e. their brightness decreases steadily from their central jets towards their outer lobes.
- They have radio luminosities of $L_{1.4\,\mathrm{GHz}} \leq 10^{32}\ \mathrm{erg\,s^{-1}\,Hz^{-1}}$.
- They show one-sided jets close to the core.
- Their jet opening angles are $\xi = dr_j/d\ell \approx 0.1$ within the first kpc. Between 1 and 10 kpc flaring occurs, with an opening angle $dr_j/d\ell \approx 0.25\ldots 0.6$ beyond that point.
- At this transition point, the flaring is accompanied by a brightening and broadening of the jet.
- Along the inner part, the magnetic field is prevailingly parallel to the jet, while a perpendicular orientation of the magnetic field with respect to the jet dominates further out.

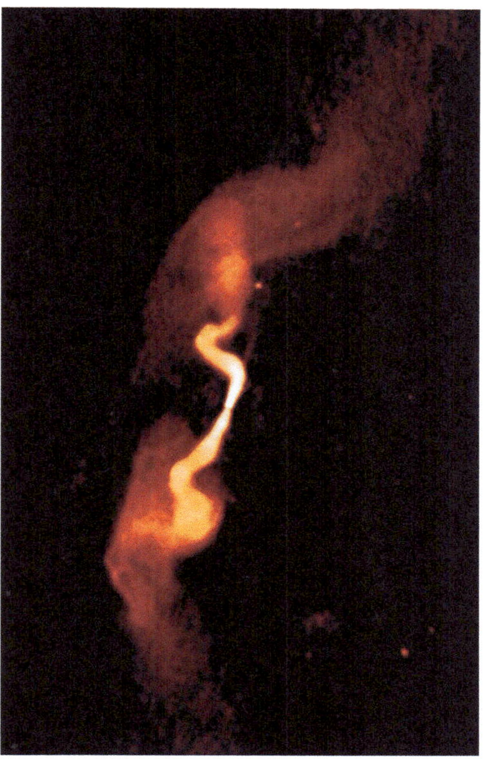

Fig. 7.9 Radio image of the FR-I radio galaxy 3C 31 (Laing et al., 2008)

- The synchrotron spectra steepen already in the diffuse lobes, with the steepening increasing further outward.

Essentially all of these properties are different from, or even opposed to, those of FR II sources. This dichotomy between FR I and FR II sources must be connected to a combination of mass infall rates and the density profile of the ISM in the host (elliptical) galaxies. A prototypical example of an FR-I radio galaxy is 3C 31, with a radio image shown in Fig. 7.9. This radio galaxy exhibits a special radio morphology with mirror symmetry about the central core, the curvature of which is interpreted in terms of the dynamical interaction of the host galaxy NGC 383 and its neighbour NGC 382 (Blandford and Icke, 1978). Exactly the same situation is seen in the FR I radio galaxy 3C 449!

A variant of FRI's is seen in the form of so-called 'head-tail radio sources', where one distinguishes further between narrow-angle tails (NAT) and wide-angle tails (WAT). The nature of these sources is the same as for 'untailed' ones, except that they occur in denser cluster environments. What happens here is that the twin jets 'drill' their channels through the ISM of the host galaxy and suddenly are exposed to the ram pressure of the surrounding intergalactic gas. From Euler's equation, we obtain

$$\rho_j \cdot v_j^2 \approx P_{ram} \cdot \frac{R}{r_j},$$

where R is the curvature radius of the jet and r_j, v_j, and ρ_j are the jet's radius, speed and density, respectively. With

$$P_{ram} = \frac{1}{2} \cdot \rho_{ICM} \cdot v_{ICM}^2$$

one has a handle on the (relative) wind speed of the ICM and may thus explore the 'cluster weather', since all other quantities can (at least roughly) be inferred from observations. It is meanwhile clear that such sources preferentially exist in unrelaxed galaxy clusters, i.e. those which just have undergone or are experiencing a merger. Such mergers give rise to large-scale turbulence, hence to strong winds inside the clusters. The particular case of the Perseus Cluster, in which the radio 'trails' of the NAT source NGC 1265 has been traced over a huge projected distance, provides very strong evidence for this. The radio tail obviously does not reflect any 'ballistic' motion of this radio galaxy in the gravitational potential of the cluster, as it is characterised by several kinks. These must be the result of intracluster turbulence deflecting the left-behind radio tails of this source (see Fig. 7.10).

This scenario is nicely discussed in an article by Burns (1998), in which he describes 'stormy cluster weather' in terms of continuous infall of dark matter and baryons in cluster mergers from their peripheries inward, this producing shocks, turbulence and winds exceeding speeds of $1000\,\mathrm{km\,s^{-1}}$ in hydrodynamic simulations, corroborated by the observed X-ray morphologies and temperatures, and their gradients.

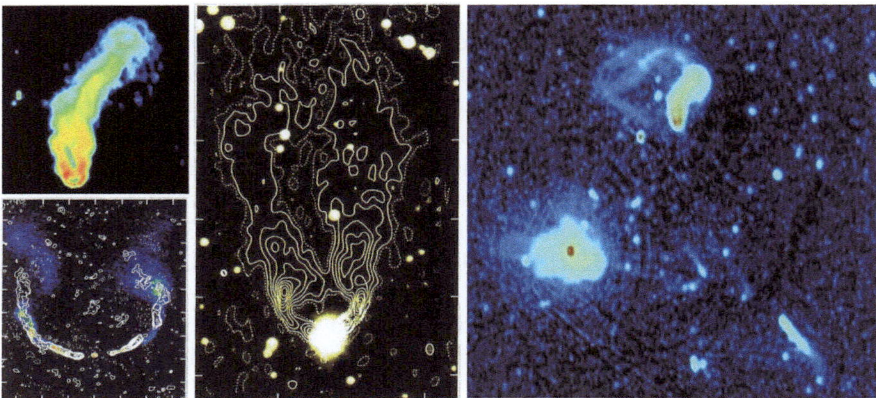

Fig. 7.10 The head-tail radio galaxy NGC 1265 at different frequencies and angular resolutions. *Top left*: 1.4 GHz (Miley et al., 1972). *Bottom left*: 5 GHz (Owen et al., 1978). *Centre*: 1.4 GHz (Wellington et al., 1973). On the right a radio image at 610 MHz of the central Perseus Cluster is shown, with NGC 1265 in the north, NGC 1275 (3C 84) in the east, and a few more narrow-angle tail sources in the south-western area (Sijbring and de Bruyn, 1998)

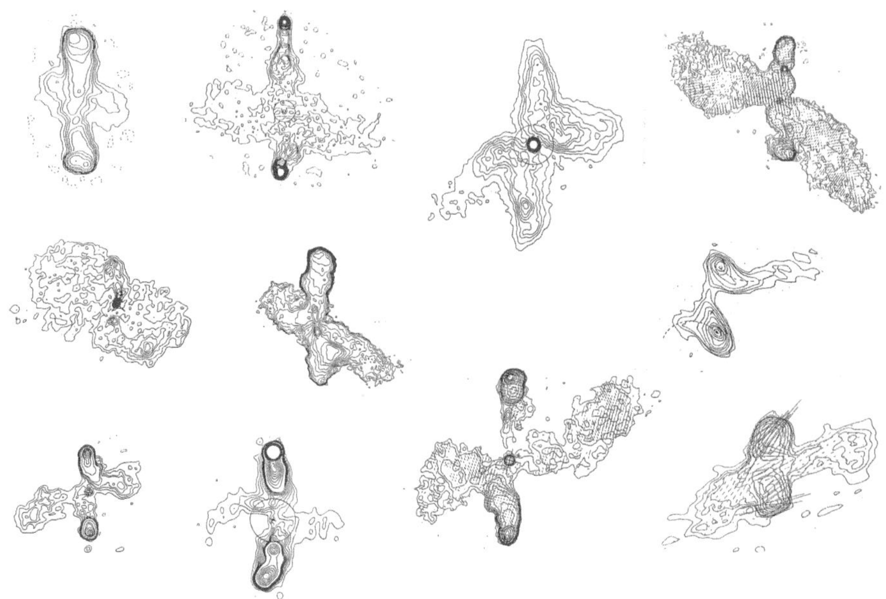

Fig. 7.11 Potpourri of so-called X-shaped radio galaxies

Another variant are the so-called X-shaped radio galaxies (Fig. 7.11). Their morphology is best explained in terms of supermassive binary black holes, which cause precession of the accretion disk of the AGN showing the jets. This leads to the X-shape, consisting of young double lobes, which are still being energised by the jets, while the aged lobes exhibit radio continuum spectra with low break frequencies.

In Sect. 8.7 we will discuss the interaction of radio galaxies with the ICM/IGM, and discuss the feedback of AGN onto these media. In particular, it turns out that the observed distribution of rotation measures across the radio galaxies provide valuable information about the surroundings of radio galaxies, rather than being intrinsic to them. Hence, radio galaxies constitute invaluable tools to study the physical parameters of the magneto-ionic medium in galaxy clusters.

7.4.3 Particle Ageing

Quite naturally, the magnetic fields in radio galaxies and the CMB photon field around them lead to synchrotron and inverse-Compton losses of the radiating particles, which is reflected in a spectral steepening of the synchrotron spectra. In FR-II sources, this steepening starts away from the hotspots, and becomes progressively stronger further away from them, towards the host galaxy, as the relativistic plasma is slowly 'falling back' from the hotspots onto the peripheral

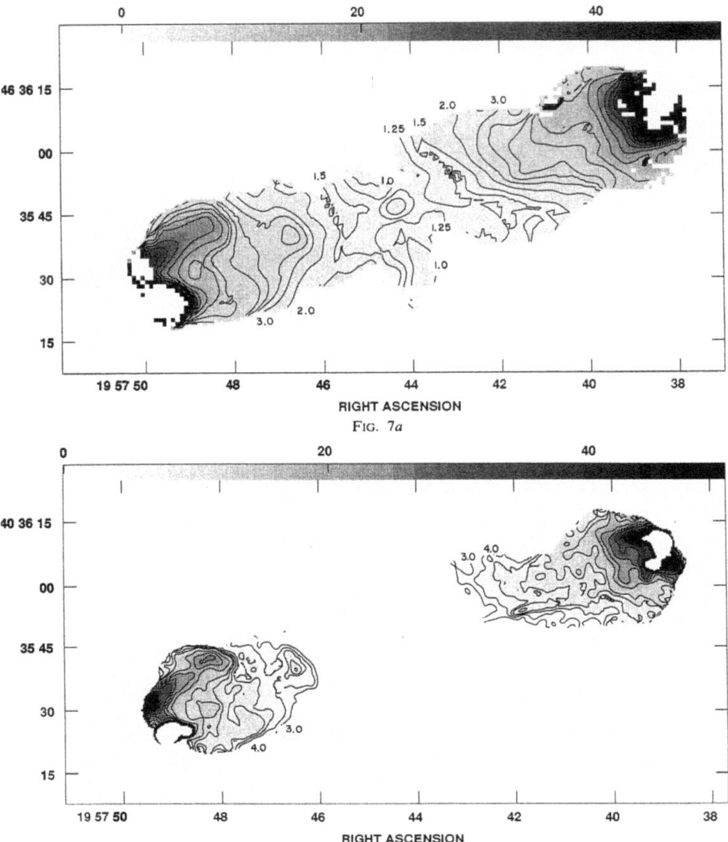

Fig. 7.12 Particle ageing as derived for Cyg A, with the break frequencies in GHz (*top*) and the ages in Myr (*bottom*) shown (Carilli et al., 1991)

ISM of that galaxy. These radiation losses in particular give rise to a break in the synchrotron spectra (see Fig. 7.12 as an example), given by

$$\nu_c = 16.1 \cdot \left(\frac{B}{\mu\mathrm{G}}\right) \cdot \left(\frac{E}{\mathrm{GeV}}\right)^2 \ \mathrm{MHz}.$$

The analysis of the synchrotron spectra in radio galaxies is a lot easier than in normal spirals: in the latter, there is contamination by thermal free-free radiation, which is totally absent in the former. It should be noted here that there are two other loss mechanisms that may play a role here: (i) Inverse-Compton losses take over as soon as the magnetic-field strength drops below $B_{CMB} = 3.25 \cdot (1 + z)\ \mu\mathrm{G}$. (ii) Adiabatic-expansion losses are dominant in the overpressured hot-spots. Assuming that most of the synchrotron radiation is radiated at the critical frequency

$$v_c = \frac{3}{4\pi} \cdot \frac{e \cdot B_\perp}{m_0 c} \cdot \gamma_{max}^2$$

the radio continuum spectra should first of all exhibit a power-law of the form $I_\nu \propto \nu^{-\alpha_{inj}}$ once a power-law energy spectrum has been established by the acceleration mechanism. The synchrotron spectrum away from the site of acceleration can be calculated by making reasonable assumptions about the time evolution of the particle energy spectrum, and integrate over this spectrum, multiplied by the frequency-dependent power, as outlined in Sect. 2.3.2. Synchrotron and inverse-Compton losses change the electron energy distribution over time at a rate

$$\frac{dE}{dt} = -\xi \cdot E^2 = -(\xi_{syn} + \xi_{IC}) \cdot E^2 \; ,$$

where the quantities ξ_{syn} and ξ_{IC} can basically be looked up in Sect. 2.4. We may, however, discern two different forms of ξ_{syn}: the so-called Jaffe-Perola (JP) model (Jaffe and Perola, 1973) takes into account that the pitch angles of the synchrotron-emitting electrons are continuously isotropised on a timescale shorter than the radiative timescale, as is also assumed in Sect. 2.4. This is probably the physically more realistic scenario, since the pitch angles of the radiating particles are permanently changed, also due to scattering off the self-induced Alfvén waves that they produce. For the JP model the resulting quantity ξ_{syn} reads ($\beta \approx 1$):

$$\xi_{syn} = \frac{4}{9} \cdot \frac{e^4}{m_0^4 c^7} B^2 \; .$$

In contrast, the Kardashev-Pacholczyk (KP) model (Kardashev, 1962; Pacholczyk, 1970) assumes that the electrons maintain their original pitch angles. In this case,

$$\xi_{syn} = \frac{2}{3} \cdot \frac{e^4}{m_0^4 c^7} \sin^2 \chi \, B^2 \; .$$

Both, the JP and the KP model assume a single (impulsive) injection of particles at time $t = 0$, at which the energy spectrum is $N(E) = N_0 E^{-g}$, while at some later time t it becomes

$$E(x_\lambda) = \begin{cases} N_0 E^{-g_{inj}}(1 - \xi E t) E^{g_{inj}-2} & \text{for } E \leq 1/\xi t \\ 0 & \text{for } E > 1/\xi t \; , \end{cases}$$

where

$$g_{inj} = 2\alpha_{inj} + 1 \; .$$

With this equation for $N(E, t)$ the synchrotron spectrum can be computed by calculating the integral (2.25). As a result, the energy losses (both for the JP and KP

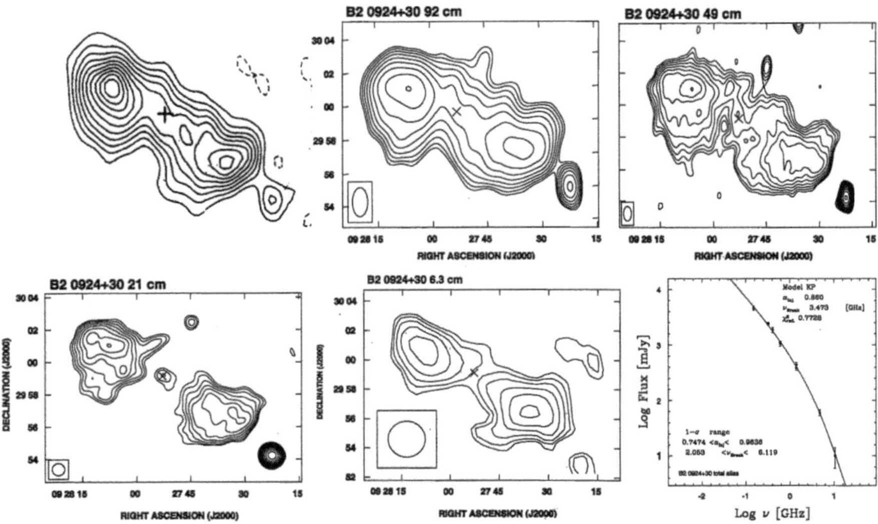

Fig. 7.13 The died radio galaxy 0924+30 at 151, 327, 610, 1400, and 4750 MHz (Jamrozy et al., 2004). The continuum spectrum in the *lower right* clearly shows the spectral break

models) cause the radio spectrum to steepen above a break frequency as was already outlined in Sect. 2.3.4. It thereby turns out that JP-spectra have an exponential cutoff, while KP-spectra steepen to a power-law with a slope $(4/3)\,\alpha_{inj} + 1$ above the break frequency, since there are always some high-energy electrons with small pitch angles that radiate at high frequencies in this latter case. In case of continuous injection (CI) of fresh relativistic particles, the spectral shapes are modified, again with a distinction to be made between constant pitch angles and isotropisation. A nice and comprehensive treatment of this was presented by Komissarov and Gubanov (1994).

In quite a few cases, the central activity of QSOs or radio galaxies has ceased, the jets have been 'switched off', and the lobes are expanding, with their synchrotron spectra steepening (i.e. the break frequency is migrating towards lower and lower values). The radio morphology of such sources is referred to as 'fat doubles', or radio relics in extreme cases. A nice example of such an aged, 'fat double', is shown in Fig. 2.21. Being one of the first 'died' radio galaxies found, 0924+30 has an 'age' of $\sim$50 Myr (Fig. 7.13).

7.5   Quasars

Quasars are the most luminous objects in the universe. Their formation commenced early-on, the most distant QSO known to date having a redshift of $z = 7.1$ (770 million years after the Big Bang). The redshift distribution shows a maximum just below $z \approx 2$, with a pronounced truncation of QSO formation below $z \approx 0.4$. QSOs reflects the history of major mergers in the universe, leading to supermassive black holes (SMBH), with the gravitational energy representing the biggest reservoir for energy production, it is these objects that are able to produce the highest luminosities via infall of matter onto accretion disks surrounding the SMBHs. Local templates of QSOs in the process of formation are so-called ultra-luminous infrared galaxies (ULIRGs), such as Arp 220 and NGC 6240. The enormous luminosity of QSOs is due to friction in their accretion disks, which heats these up to $T_{ad} \leq 10^8$ K. This friction arises from the fact that conservation of angular momentum does not allow matter to fall onto or into the black hole straight, but rather 'spirals' inward in an accretion disk, which is subject to very strong differential rotation.

At a redshift of $z \approx 3$, the feeding rate onto QSO SMBHs was about 300 times that of what is found in the local universe. Most giant elliptical galaxies can be considered as 'dead quasars' – the relics of a period with stronger merger and hence QSO activity. Not all QSOs are also strong radio (i.e. synchrotron) sources. In fact, there are many more 'radio-quiet' than 'radio-loud' quasars. This is quantified by the R-parameter, which is the ratio of optical-to-radio flux density, $R = S(4400\,\text{Å})/S(5\,\text{GHz})$, and $R = 10$ being defined as the dividing line between radio-quiet and radio-loud. Recent MHD simulations of rotating black holes with accretion disks seem to indicate that the spin of the central SMBH may be the underlying reason for this dichotomy.

Radio-loud quasars show one-sided jets, which sometimes also produce large-scale lobes (Fig. 7.14). Their jets can be traced down to pc-scales by means of (space-) VLBI. Monitoring the jet components[1] discloses changes in the jet morphologies, with multiple components being traced over decades (Lister et al., 2009). It is thus possible to construct kinematic models of the ejection and propagation of the relativistic plasma.

Quite a bit of information has recently been gained by measurements of the linear polarisation and rotation-measure studies of the central regions of QSOs with the VLBA. However, the interpretation of observed rotation measures in the central regions of AGN may not be straight forward. There are several possible effects that could mimic Faraday rotation where there is none. High-resolution VLBI observations reveal an unresolved core in quasars, plus jet components that have recently emerged from that core region. These components might exhibit different polarisation angles, owing to

  (i) Different intrinsic polarisation angles,
 (ii) Relativistic aberration of subcomponents having identical polarisation angles but different relativistic speeds,
(iii) A rotation-measure gradient across the core.

---

[1] MOJAVE: Monitoring of Jets in AGN with VLBA Experiments.

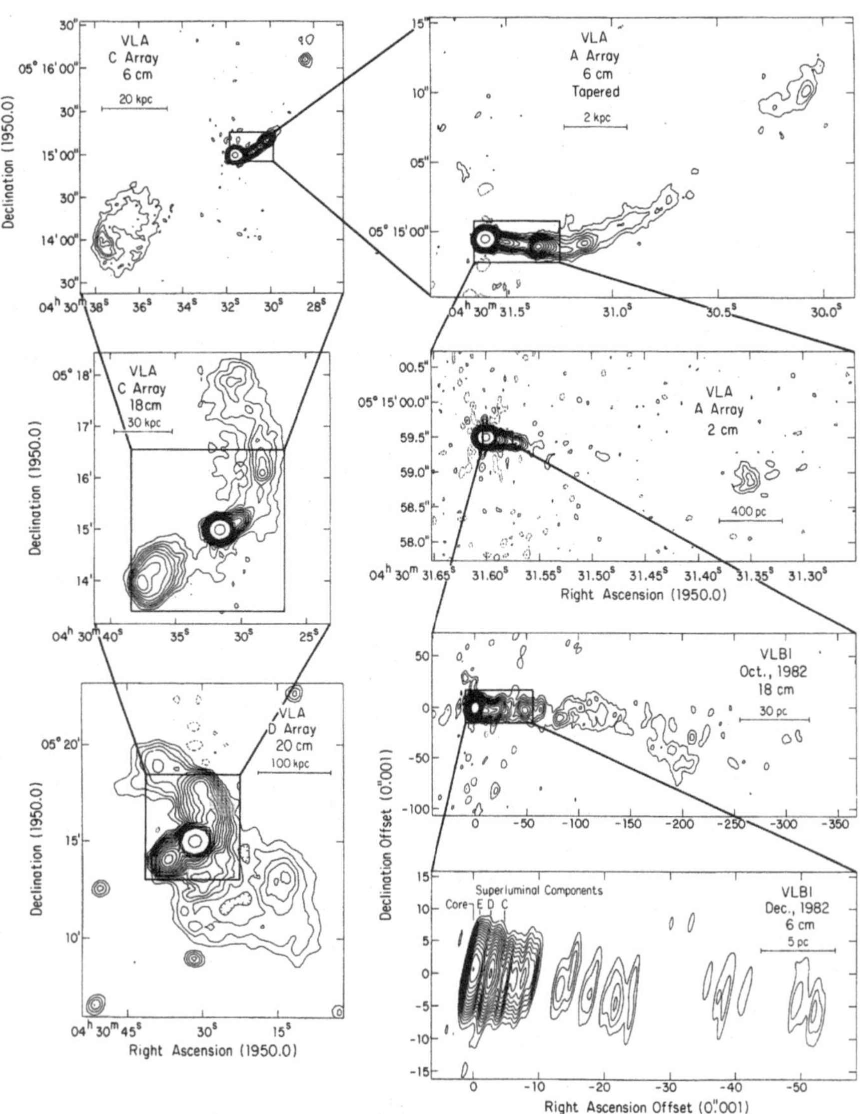

Fig. 7.14 The quasar 3C 120 at different wavelengths and angular resolutions (Walker et al., 1987)

If the synchrotron opacity of the components is different or if they have different spectral indices, this could change their relative preponderance, which would result in dramatic changes of the polarisation angle as a function of frequency.

Recent VLBA observations of the quasars 3C 273, 3C 279, and 3C 380 have revealed the following interesting properties (see also Fig. 7.15): the central cores

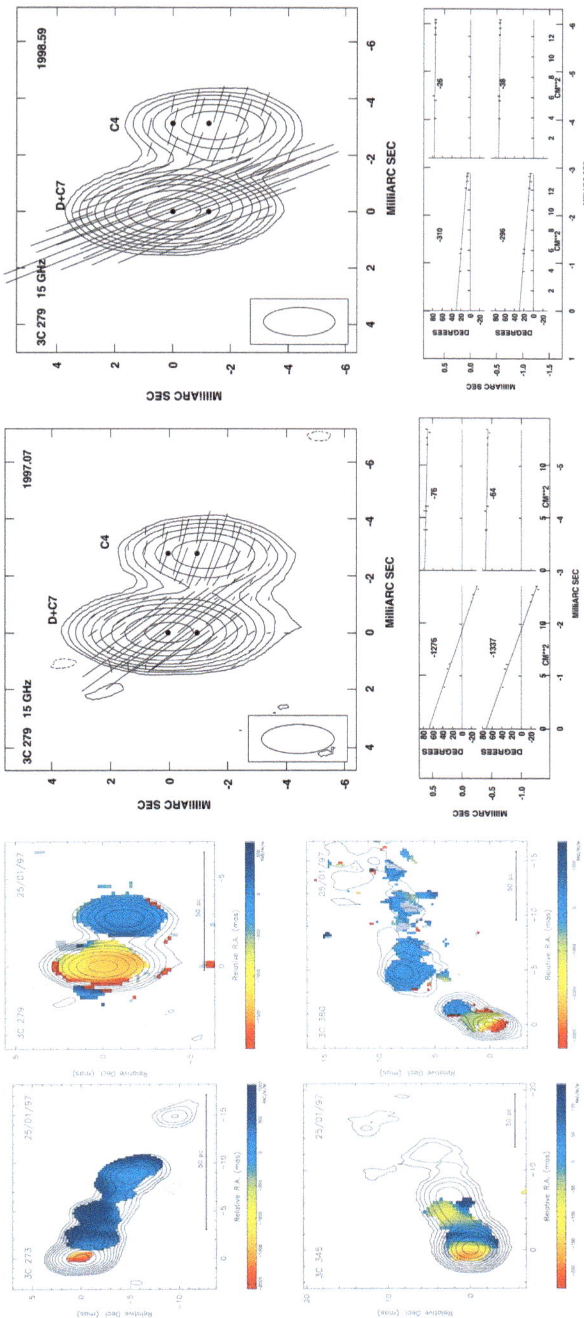

Fig. 7.15 Rotation measures in the central regions of quasars (Taylor, 1998; 2000)

have rotation measures of $RM > 1000\,\mathrm{rad\,m^{-2}}$. Beyond about $20\,\mathrm{pc}\;h_{75}^{-1}$ ($h_{75} = H_0/75\,\mathrm{km\,s^{-1}\,Mpc^{-1}}$), they show rotation measures of $RM < 100\,\mathrm{rad\,m^{-2}}$. Such sharp rotation-measure gradients cannot be produced by magnetic fields in the foreground, i.e. either in galaxy clusters or in the Milky Way. They must be produced by ordered magnetic fields in the central regions of these QSOs on scales of $1\dots100\,\mathrm{pc}$. If associated with the NLR,[2] then $B_u \approx 0.05$ G on a 10-pc scale is inferred from the observed rotation measures. The high rotation measures found in these QSO cores account for the depolarisation found at cm wavelengths. They result from irregularities in the Faraday screens on scales smaller than the telescope beam.

7.6   Seyfert Galaxies

As mentioned before, Seyfert galaxies are mostly spiral galaxies hosting an AGN in their centres. Depending on the viewing angle, the observer sees a Seyfert-1 galaxy with broad emission lines, or a Seyfert-2 type with narrow emission lines. What is naturally common to all Seyfert galaxies is that they stick out of the radio-FIR correlation (Fig. 7.1), in the sense that they are radio-overluminous compared to normal star-forming galaxies. The radio excess is not as large as in case of radio galaxies or QSOs, but since this correlation is so tight for normal galaxies, it can be used to pick out most active spirals right away.

Most Seyfert galaxies host compact sources of synchrotron radiation produced by the central AGN. However, there are quite a few examples in which the spiral disk galaxies have produced extended radio lobes – in some cases having very much the same morphology as classical radio lobes produced by radio galaxies (see Fig. 5.6). Apparently, in these cases the orientation of the jets is such that they avoid the densest ISM and make it out of the gaseous disk. Here, we briefly feature these objects by describing their radio continuum properties in tabular form. Note that all sizes are projected sizes, so the observed structures could still be larger.

---

[2] The BLR can be ruled out: it is too dense, hence it must be depolarised.

Galaxy/object	Morphology	Jet orientation	$\vec{B}$-field
0313-192, QSO/Blazar?	350 kpc double source with jets	30° off the poles	Not known
0400-181	Jets & double lobes; size unknown (no distance known)	Not known	Not known
0412+040 Seyfert 2	90 kpc; core & double lobes	Not known	Not known
IC 2497 LINER[a]	Extended radio continuum $\perp$ to disk; likely to be an exhausted AGN	$\sim \perp$ disk,	Not known
NGC 1068 Seyfert 2	1-kpc lobes; nucleus, inner 'hot-spots'	Slightly inclined out of the disk	Similar to that in classical lobes; at lobes' head: $B_t \sim 400 - 600\,\mu G$
NGC 3367 Seyfert-like	12 kpc; double lobes, S-shaped $\Rightarrow$ interaction with ambient gas?	Slightly inclined out of the disk (poln. asymmetry)	Not known (Faraday rotation)
NGC 4258 Seyfert/LINER	14 kpc; "anomalous radio arms", S-shaped; bifurcation	Most likely in plane (from RM)	$\parallel$ to jet

[a]A low-ionisation nuclear emission-line region (LINER) is a galactic nucleus showing spectra that typically include line emission from weakly ionised or neutral atoms, such as O, O^+, N^{++}, and S^+, with emission from strongly ionised atoms, such as O^{++}, Ne^{++}, and He^+ being relatively weak

7.6.1 *Jet Formation*

An inevitable question is how the jets come into existence. It was speculated early-on that it is the magnetic field near to the accretion disk that plays a cardinal role in jet formation. Meanwhile, a large number of MHD simulations has been carried out (an example shown in Fig. 7.16), which in essence reflect the response of the magnetic field to the rotation of the accretion disk. Since the field is tightly coupled to that disk, its originally poloidal structure is quickly deformed, with more and more torodial components being generated, until the field is strongly wound up. The plasma is now accelerated in this torodial field by what can be interpreted as a centrifugal force in a co-rotating frame ($\vec{j} \times \vec{B}$ forces). In the end, this force is due to the magnetic pressure gradient.

Hydrodynamic numerical simulations have been carried out quite successfully, these allowing to deduce the physical conditions in the ICM or IGM into which the jets and lobes of radio galaxies are expanding, i.e. the pressure, temperature,

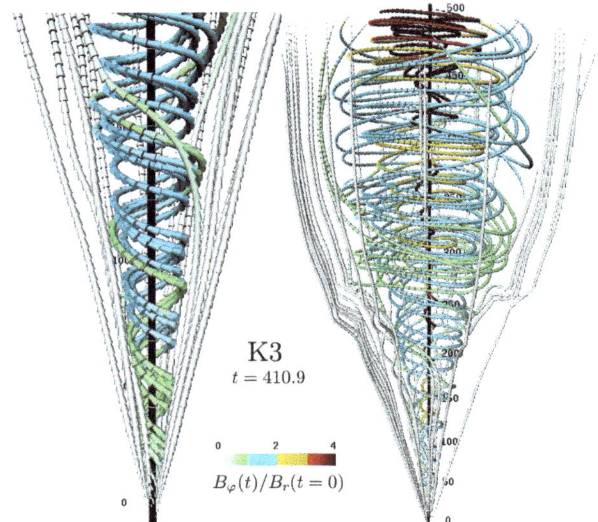

Fig. 7.16 MHD simulations of a jet (Moll et al., 2008)

and magnetic-field strength and structure. Apart from getting a clue how the jets are produced via twisting of the magnetic fields, these were found to have two important effects on the shape and evolution of radio galaxies:

(i) Magnetic fields damp Kelvin-Helmholtz instabilities at the interface between the cocoon and the ambient gas (that of the IGM), this stabilizing the contact surface. This way, pronounced jet heads and lobes such as seen in Cyg A and other FR IIs are produced.
(ii) When including magnetic fields in the simulations, the amount of entrainment of the ambient gas is much lower than without magnetic fields.

7.6.2   *Jet Composition*

It is hitherto not clear what the jets consist of. For sure, they possess a magnetic field, otherwise we would not see (polarised) synchrotron radiation. But what kind of relativistic particles produces this synchrotron radiation? The could be 'heavy' or 'light' one, i.e. we could be dealing with

(i) A mildly relativistic $p - e^-$ jet, with a jet speed of $v_j \leq 0.5 \cdot c$. It would carry most of the mass and kinetic power ejected by the AGN, and would be responsible for the formation of the kpc-scale jets, hot-spots, and lobes;
(ii) A relativistic $e^+ - e^-$ beam ("pair plasma"), with a jet speed of $v_b \approx c$ and $3 \leq \gamma_b \leq 10$. This jet material would move in a channel through the jet and would be responsible for the superluminal motion of, and γ-ray emission seen from, the jets.

There are three lines of argument that may shed light onto the question of jet composition (see, e.g. Wardle et al. 1998). The first one is simply that the total energy dissipated by the jets in the lobes in the form of magnetic fields and relativistic particles must match the total energy flux carried by the jets. In Sect. 2.3.2 we saw that the energy distribution of the relativistic particles follow a power law, $N(E) \propto E^{-g} \propto \gamma^{-2\alpha+1}$. If, with typical values of the spectral index α the Lorentz factor γ extends to low values, $\gamma \lesssim 100$, then the total energy flux of a $p - e^-$ jet would be exceedingly large as compared to the dissipated energy one measures in the lobes. For an $e^+ - e^-$ jet the requirement is exactly opposite: the energy spectrum would have to extend to very low energies, $\gamma \approx 1$ in order to match the observed dissipation. Hence, if one would find $\gamma \ll 100$, this would be strong evidence in favour of a pair-plasma jet. However, such mildly relativistic particles can not be observed, as their radiation is strongly self-absorbed. Such a low-energy cutoff might be visible at X-ray or γ-ray energies via inverse-Compton radiation, but this is a real observational challenge. The second argument is based upon the observed linear polarisation that is commonly observed in the jets, limiting the amount of internal Faraday rotation, which is mainly produced by the lowest-energy electrons in the jet. For an $p - e^-$ jet, the resulting Lorentz factor turns out to be $\gamma \approx 100$. An $e^+ - e^-$ jet obviously can not produce any internal Faraday rotation, since electrons and positrons gyrate in opposite directions. The third observational evidence may come from the measurement of circular polarisation of jets. It is produced by Faraday conversion (see, e.g., Beckert and Falcke, 2002) of linear to circular polarisation, which is a propagation effect and necessitates the radiating particles to have Lorentz factors $\gamma \ll 100$, implying an $e^+ - e^-$ jet. A detailed review of the circular polarisation in AGN was given by Macquart and Fender (2004).

A pair plasma could be inferred from detections of the 511-keV emission annihilation line from jets. If the jet plasma mixes with the surrounding gas, this will thermalise the positrons in the jet, leading to continuous annihilation with ambient electrons. However, no detection of the line has been made up to now (e.g. Marscher et al., 2007).

7.7 Origin of Magnetic Fields

The origin of magnetic fields in AGN is still a matter of speculation. The easiest path of arguing is probably connected with the rotational energy of the central black hole. Rees (2006) takes an optimistic view in saying that the time scales for any relevant processes are extremely short close to a black hole – maybe as short as hours, so that there has always been enough time for a battery process to operate, or for a dynamo to be seeded by an infinitesimal field.

Following Hoyle (1969), one can give an order of magnitude for the possible field strengths in AGN. The rotational energy of a compact central object from which the magnetic energy is tapped can be parametrised as

$$E_{rot} = f \, M_{\bullet} \, c^2,$$

where $M_{\bullet}$ is the mass of the compact object and the efficiency factor $f < 1$. If there is equipartition between the magnetic field and the rotational energy in the fluid (achieved by differential rotation and/or by the dynamo action) this implies a field strength of

$$B_c \approx \left(\frac{8 \, \pi \, f \, M_{\bullet} \, c^2}{V_c} \right)^{\frac{1}{2}},$$

where V_c is the volume of the central region. The accretion disks offer optimum conditions to drive a dynamo. For a region of size ~ 1 pc, we then have

$$B_c = 0.6 \cdot \left(\frac{M_{\bullet}}{M_{\odot}} \right)^{\frac{1}{2}} \cdot f^{\frac{1}{2}} \ \mathrm{G}.$$

References

Baade, W., Minkowski, R., 1954, Astroph. J. **119**, 206
Beckert, T., Falcke, H., 2002, Astron. Astroph. **388**, 1106
Blandford, R.D., Icke, V., 1978, MNRAS **185**, 527
Bolton, J.G., Stanley, G.J., 1948, Nature **161**, 312
Burns, J.A., 1998, Science **280**, 345
Carilli, C.L., Perley, R.A., Dreher, J.W., Leahy, J.P., 1991, Astroph. J. **383**, 554
Carilli, C.L., Harris, D.E. (edts.), 1996, *Cygnus A: Study of a Radio Galaxy*, Cambridge: CUP
Dreher, J. W.; Carilli, C. L.; Perley, R. A., 1987, Astroph. J. **316**, 611
Fanaroff, B.L., Riley, J.M., 1974, MNRAS **167**, 31
Hoyle, F., 1969, Nature **223**, 936
Jamrozy, M., Klein, U., Mack, K.-H., et al., 2004, Astron. Astroph. **427**, 79
Jaffe, W.J., Perola, G.C., 1973, Astron. Astroph. **26**, 423
Kardashev, N.S., 1962, Sov. Astron. **6**, 317
Komissarov, S.S., Gubanov, A.G., 1994, Astron. Astroph. **285**, 27
Laing, R.A., Bridle, A.H., Parma, P., et al., 2008, MNRAS **386**, 657
Lister, M.L., Aller, H.D., Aller, M.F., et al., 2009, Astron. J. **137**, 3718
Macquart, J.-P., Fender, R.P., (edts.), 2004, *Circular Polarisation from Relativistic Jet Sources*, Kluwer Academic Publishers, Dordrecht
Marscher, A.P., Jorstad, S.G., Gómez, J., et al., 2007, Astroph. **665**, 232
Matthews, T.A., Sandage, A.R., Astroph. J. **138**, 30
Miley, G.K., Perola, G.C., van der Kruit, P.C., et al., 1972, Nature **237**, 269
Moll, R., Spruit, H.C., Obergaulinger, M., 2008, Astron. Astroph. **492**, 621
Owen, F.N., Burns, J.O., Rudnick, L., 1978, Astroph. J. **226**, L119
Pacholczyk, A.G., 1970, *Radio astrophysics. Nonthermal processes in galactic and extragalactic sources*, Series of Books in Astronomy and Astrophysics (San Francisco: Freeman)
Perley, R.A., Dreher, J.W., Cowan, J.J., 1984a, Astroph. J. **285**, 35
Perley, R.A., Bridle, A.H., Willis, A.G., 1984b, Astroph. J. Suppl. **54**, 291
Rees, M.J., 2006, Astron. Nachr. **327**, 395
Russell, C.T., Elphic, R.C., 1979, Nature **279**, 616

Schmidt, M., 1963, Nature **197**, 1040
Sijbring, D., de Bruyn, A.G., 1998, Astron. Astroph. **331**, 901
Smith, F.G., 1951, Nature **168**, 555
Seyfert, C.K., 1943, Astroph. J. **97**, 28
Sopp, H.M., Alexander, P., 1991, MNRAS **251**, 145
Taylor, G.B., 1998, Astroph. J. **506**, 637
Taylor, G.B., 2000, Astroph. J. **533**, 95
Urry, C.M., Padovani, P., 1995, PASP **107**, 803
Walker, R.C., Benson, J.M., Unwin, S.C., 1987, Astroph. J. **316**, 546
Wardle, J.F.C., Homan, D.C., Ojha, R., Roberts, D.H., 1998, Nature **395**, 457
Wellington, K.J., Miley, G.K., van der Laan, H., 1973, Nature **244**, 502

Chapter 8
Intergalactic Magnetic Fields

8.1 Clusters of Galaxies

Clusters of galaxies are the largest gravitationally bound entities in the universe. They comprise up to several thousand galaxies, the majority of which are dwarf galaxies. For example, the Virgo Cluster contains some 250 large galaxies and more than 2000 smaller ones. Small assemblies of galaxies are referred to as groups. For instance, the local group comprises two really massive galaxies, the Milky Way and M 31, and a third one, M 33, being somewhat less massive, followed by the Large and Small Magellanic Cloud and some 30 other dwarf galaxies with masses down to what is comparable to the largest globular clusters.

An important ingredient of galaxy clusters is the ICM (or IGM), which is hot ($T \simeq 10^8$ K), tenuous ($n_e \approx 10^{-3}$ cm^{-3}) gas, and which is well 'visible' through its bright X-ray emission ($L_X \approx 10^{43} \ldots 10^{45}$ erg s^{-1}), produced by thermal bremsstrahlung. Another constituent of the ICM is its relativistic plasma, consisting of relativistic particles and magnetic fields – the topic of this textbook, which we shall come back to. The cluster mass is, as is well-known, dominated by dark matter, as inferred from the galaxies' velocity dispersion, from the temperature of the hot gas, and from gravitational lensing. In the framework of the hierarchical scenario of galaxy cluster formation in a ΛCDM cosmology, large clusters have formed by gravitational mergers from smaller entities, i.e. from groups or subclusters. In these mergers, high kinetic energies are involved, viz. up to 10^{64} erg. A large fraction of this energy is dissipated in the ICM in the form of shocks, turbulence and bulk motion, which heat up the gas.

Many clusters exhibit strongly peaked X-ray surface brightnesses in their centres, implying high densities, and cooling times within the inner 100 kpc of $t_{cool} \ll t_{Hubble}$. Hence, in order to maintain hydrostatic equilibrium, inward flows may be required. X-ray observations with XMM-Newton, which has a high spectral resolution, did not indicate any evidence for large amounts of cooling and condensing gas in the centres of galaxy clusters. It thus seems that such cooling flows' are hindered

© Springer International Publishing Switzerland 2015
U. Klein, A. Fletcher, *Galactic and Intergalactic Magnetic Fields*,
Springer Praxis Books, DOI 10.1007/978-3-319-08942-3_8

by some as yet unknown mechanism. There is consensus that 'cooling core' clusters
are dynamically more relaxed than non-cooling core ones.

A full understanding of the ICM necessitates adequate knowledge of the role of
the nonthermal component, the best information about which is obviously provided
by radio continuum observations. A very strong argument comes from the fact that
quite a number of clusters exhibits diffuse radio sources that cannot be associated
with any discrete sources powering them, but are rather connected with the general
ICM. The observed synchrotron emission requires $\sim$GeV electrons, hence highly
relativistic particles, and magnetic fields at the level of μG strengths, which must
pervade – at least part of – the cluster volumes (for an observational review,
see Feretti et al., 2012). Independent evidence for such magnetic fields comes
from rotation measures of polarised background sources (Clarke et al., 2001) or
radio galaxies located within these clusters (Bonafede, 2010). Finally, independent
evidence for such a population of relativistic particles also comes from nonthermal
emission of inverse-Compton origin seen in the regime of hard X-rays and possibly
in the EUV. The combination of the diffuse radio continuum emission with the hard
X-ray radiation can be used to estimate intra-cluster magnetic-field strengths.

Note that the non-thermal radio emission from cosmic-ray electrons in the
vicinity of merging galaxy clusters is an important tracer of ongoing or past cluster
mergers. Together with the observed soft X-ray emission, it can shed light on the
dynamical state of a galaxy cluster.

8.2 Radio Haloes

Cluster radio haloes are the most spectacular manifestation of diffuse nonthermal
emission and widespread magnetic fields seen over large portions of galaxy clusters.
They permeate them on scales of more than a Mpc, with the following properties:

- They have intensities of $I_{1.4\,\mathrm{GHz}} \approx 10^{-6}\,\mathrm{Jy}/\square''$.
- Their radio spectral index is $\alpha \geq 1$ ($I_\nu \propto \nu^{-\alpha}$).
- They are unpolarised down to the current detection levels for polarised radio
  synchrotron emission.

For comparison, the mean radio brightness of the radio lobes of the classical
radio galaxy Cyg A is of order $0.5\,\mathrm{Jy}/\square''$, hence 500 000 times more intense.[1] The
first such diffuse cluster radio source was reported by Willson (1970), and this
structure was later on mapped with ever increasing resolution and sensitivity (see,
e.g. Fig. 8.1). Discovered in the Coma Cluster of galaxies (A 1656), this structure
soon showed a spectral steepening from the centre towards the periphery, with
$\alpha = 0.8$ in the centre to $\alpha = 1.8$, and $\langle \alpha \rangle = 1.3$. The break frequency, reflecting
the synchrotron age of the radiating particles, was determined as $\nu_c \geq 1\,\mathrm{GHz}$, and

---

[1] Recall that brightness is distance-independent!

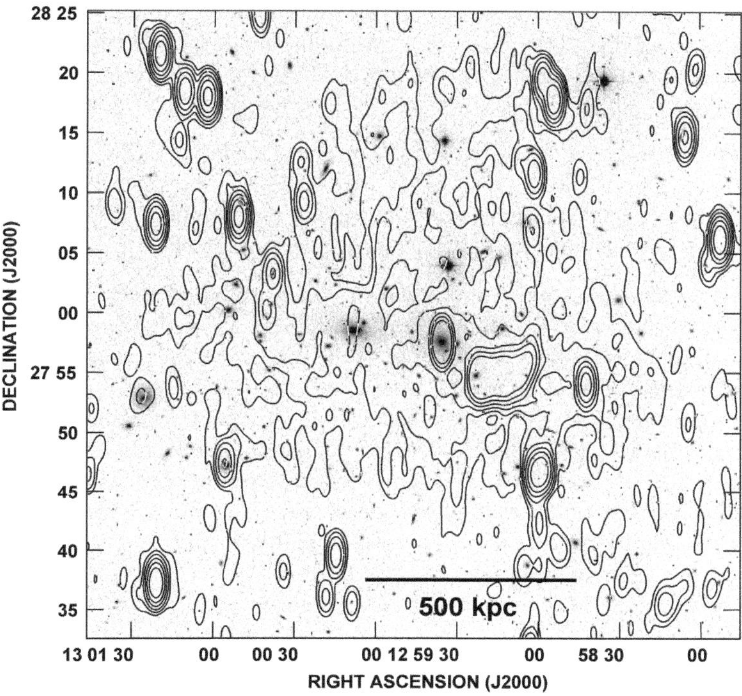

Fig. 8.1 The Coma Cluster at $\lambda = 90$ cm, superimposed onto an optical image from the POSS (Feretti et al., 2012)

the average total magnetic-field strength as $B_t \approx 0.6 \, \mu$G or, more precisely (the equipartition strength)

$$B_{eq} = 0.57 \cdot (1 + \beta)^{0.26} \, \mu\text{G},$$

where β is the ratio of the energy contained by the protons to that contained by the electrons (Sect. 3.2). It is a common feature of all clusters with radio haloes that they are X-ray luminous. Such radio haloes have been found in about one quarter of all clusters with $L_X > 5 \cdot 10^{44}$ erg s^{-1}. Studies of several other cluster haloes have been performed, such as A 2163, A 2256, 1E 0657-57, and distant clusters like A 2744 ($z = 0.308$) and CL 0016716 ($z = 0.5545$), the latter being the most distant cluster with a radio halo known to date. Smaller-sized haloes (500 . . . 600 kpc) have been detected in a few other cases (e.g. A 2218, A 3562). The total number of radio haloes discovered at the time of writing this is about 50.

A common property of cluster radio haloes is their lack of any strong (linear) polarisation, even though the presence of central synchrotron haloes implies the existence of magnetic fields at least throughout the central volumes of galaxy clusters. The degree of polarisation is $<10\%$ (Coma), or even $<5\%$ in others, obtained at 1.4 GHz. Obviously, any polarised radiation emitted by these central radio haloes traverses large lines of sight through the cluster medium. This, along with the arguments given below, delivers the likely reasons for the low degrees of polarisation:

- Internal depolarisation, owing to mixing of the magnetic fields with the thermal gas.
- Turbulence causing highly disordered fields, which could come along with a likely mechanism that accelerates the particles, viz. galactic wakes.

Probably the only way to shed light on this unresolved question is to employ the newly developed RM synthesis technique (Sect. 3.3.8) and to make use of the improved angular resolution of future radio interferometers such as the *Square Kilometer Array* (SKA). However, given the low intensities of radio haloes, this will require very high sensitivity, hence long observing times. So far, the low surface brightness of the radio haloes so far restricted any studies in most cases to frequencies of about 1.4 GHz and below, with correspondingly low angular resolution. The situation is changing while this is being written, LOFAR has become operational and has started to produce high-quality images also of galaxy clusters at the lowest radio frequencies. The SKA pathfinders (MeerKAT, ASKAP) are currently paving the way towards extreme sensitivity paired with very high angular resolution.

From the beginning of their discovery, the cardinal issue has been to clarify the origin of cluster haloes. They are not associated with any discrete sources such as radio galaxies. Even though radio galaxies are always present in galaxy clusters, their jets can not possibly produce these huge synchrotron haloes, as the time scale for particle ageing is too short to keep the ejected particles relativistic across such large volumes. As demonstrated in Sect. 7.4.2 (Fig. 7.10), radio galaxies leave behind marked trails of synchrotron radiation, which is visible for several 10^7 year. In order to explain cluster radio haloes, there are various possible models, which can basically be divided into two classes:

- Primary electron models, in which the relativistic electrons are injected into the IGM by AGN (QSOs, radio galaxies,...) and/or by starburst galaxies (SNe, galactic winds). However, the radiative lifetime of such particles is rather short,

$$t_{\frac{1}{2}} = 8.35 \cdot 10^9 \cdot \left(\frac{B}{\mu G} \right)^{-2} \cdot \left(\frac{E}{GeV} \right)^{-1} \text{ year;}$$

hence such particles would be rendered invisible after $10^7 \dots 10^8$ years, and would thus require continuous injection by sources (which is in conflict with the observations) or by some reacceleration mechanism. One possible process is turbulence caused by galactic wakes, although its efficiency is a matter of debate.

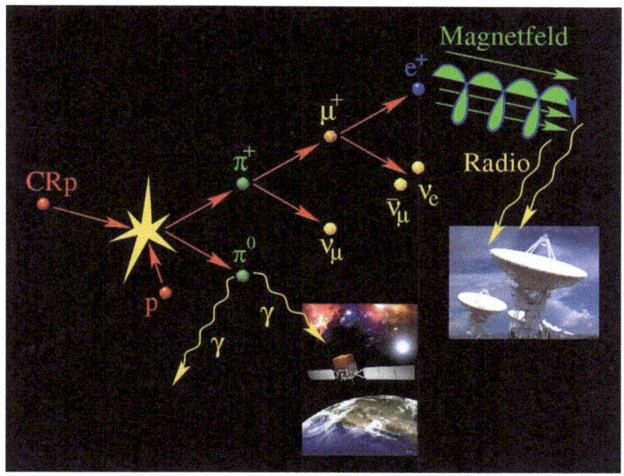

Fig. 8.2 Hadronic collisions in the ICM (Credit: C. Pfrommer)

- Secondary electron models, in which the relativistic electrons result as secondary products in hadronic collisions. Relativistic protons have lifetimes exceeding a Hubble time. They can thus propagate over large distances from their site of origin. In the central regions of clusters, they frequently collide with the thermal protons of the ICM, thereby producing relativistic e^- or e^+ and γ-rays from the π^0 decay. Future γ-ray observatories can therefore test such models independently.

The production of secondaries proceeds accordingly to the following chain (see also Fig. 8.2):

$$p + p \rightarrow \pi^{+/0} + 2\,N,$$

where $2\,N$ is any combination of particles. This is followed by a mesonic decay

$$\pi^+ \rightarrow \mu^+ + \nu_\mu,$$
$$\pi^0 \rightarrow 2\,\gamma,$$

and by the leptonic decay

$$\mu^+ \rightarrow e^+ + \nu_e + \bar{\nu}_\mu.$$

So this finally results in relativistic electrons or positrons that we could see in the form of synchrotron radiation. A correlation betwee n the radio luminosity of cluster haloes and the X-ray luminosity of their hot gas has been established, which is

$$P_{1.4\,\mathrm{GHz}} \propto L_X^{1.97 \pm 0.25}, \tag{8.1}$$

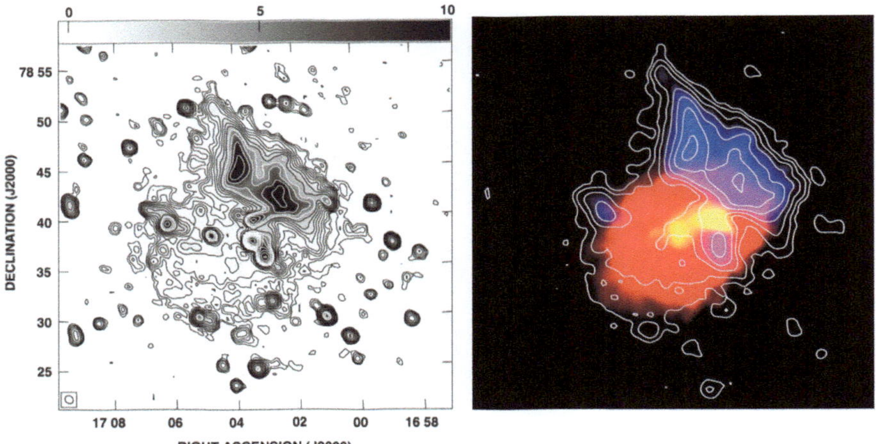

Fig. 8.3 Radio continuum images of the galaxy cluster A 2256 (Clarke and Enßlin, 2006). In the *right panel*, the radio contours have been superimposed onto an X-ray image (colour), with radio point sources subtracted

where the X-rays have been measured in the 0.1–2.4 keV energy band. This correlation obviously supports models of secondary-electron production in hadronic collisions. On the other hand, the fact that central radio haloes appear to be connected with clusters having substructure, hence are undergoing or have undergone a merger, speaks in favour of primary models, with turbulence and shocks as energy sources. Hence, this issue still awaits clarification. This must come with future γ-ray observatories (the *Fermi* Gamma-Ray Space Telescope was unable to detect γ-rays from clusters), and it must come with the LOFAR harvest, aided by numerical (MHD) simulations that are currently also underway.

As another prominent example, the radio halo of A 2256 is shown in Fig. 8.3, which has been thoroughly studied. It constitutes a perfect template for cluster radio haloes, but also exhibits a so-called radio relic, which is an extended and bright peripheral structure (see Sect. 8.4). After subtraction of the (background) point sources, the central extended radio halo is readily seen to coincide with the hot gas, while the relic is visible as a bright, extended structure in the north-west. While the relic shows degrees of polarisation of up to 45 %, no polarisation has been detected in the halo so far. A 2256 was the first galaxy cluster in which extended low-frequency (between 18 and 67 MHz) radio synchrotron emission has been detected with LOFAR (van Weeren et al. 2013). Both the radio halo and the giant relic were detected 63 MHz, and the diffuse, extended radio emission remains visible down to 20 MHz.

8.3   Rotation Measures

Faraday rotation of both, background as well as cluster-source emission is another, independent hint at the ICM being magnetised. Clarke et al. (2001) analysed the rotation measures of background sources probing 16 low-redshift ($z \leq 0.1$) galaxy clusters (Fig. 8.4), selected to not possess any radio haloes or being indicative of any cooling flows. Combined with an analysis of cluster-embedded sources, they found a cluster-generated rotation-measure excess out to $\sim$0.5 Mpc h_{75}^{-1} from the cluster centres. The results imply magnetic fields with a high filling factor and

$$\langle B \rangle = (5 \ldots 10) \cdot \left(\frac{\ell}{10\,\text{kpc}} \right)^{-\frac{1}{2}} \cdot h_{75}^{1/2}\ \mu\text{G},$$

where ℓ is the correlation length of the magnetic field. Rotation-measure mapping in clusters of galaxies has been performed in quite a number of cases using extended radio galaxies. The first such high-resolution rotation-measure study was performed using Cyg A. These results imply that a Faraday screen with $-4000\,\text{rad}\,\text{m}^{-2} < RM < +3000\,\text{rad}\,\text{m}^{-2}$ is required! The screen is implied because $\Delta\psi \sim \lambda^2$. Recall that strong deviations from this law imply mixing of Faraday-rotation with the emitting medium (Sect. 3.3.2). In line with this scenario of a foreground Faraday screen, the Faraday rotation and depolarisation is generally lower for the lobe fed by the brighter jet, which is the nearer side as seen from the observer. This phenomenon has been coined 'Laing-Garrington effect' (Garrington et al. 1988). The observed variations or gradients of the rotation measures, which are of the order of $|RM| \approx 300\,\text{rad}\,\text{m}^{-2}/\text{arcsec}$, imply size scales that are too small to be attributable to any variations in the Galactic foreground. From these observations, magnetic fields of strength $2 \ldots 10\,\mu\text{G}$, ordered over scales of $20 \ldots 30\,\text{kpc}$, are inferred.

Further detailed rotation-measure studies have been performed on what is deemed cooling-core clusters, characterised by powerful radio galaxies in their

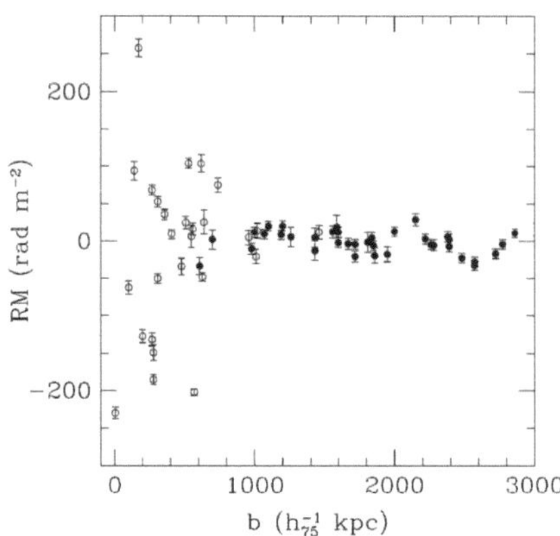

Fig. 8.4 Radial distribution of RMs in galaxy clusters (Clarke et al., 2001)

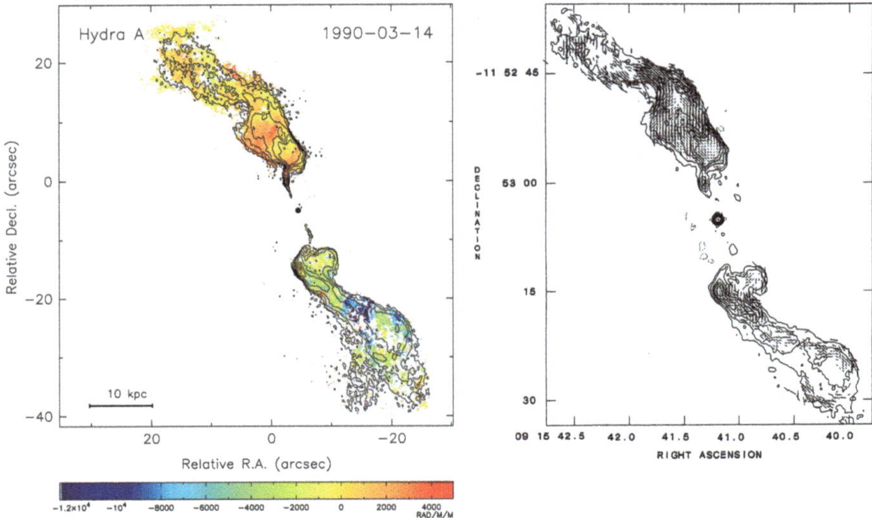

Fig. 8.5 Rotation measure (*left*) and magnetic field (*right*) in the cooling-core radio galaxy Hyd A (Taylor et al., 1990; Taylor and Perley, 1993)

centres. Like in case of Cyg A, these sources exhibit large rotation measures, with strong spatial variations. Prominent examples are Hyd A (Fig. 8.5), 3C 295, or Vir A. The inferred magnetic-field strengths range from $\sim$5 to $\sim$30 μG. The ratio of thermal to magnetic energy density, the so-called 'plasma-β',

$$\beta = \frac{n\,k\,T}{B^2/\,8\pi},$$

is found to exceed unity significantly. For instance, in 3 C 31 one finds $\beta \approx 10$ at the radius of the galaxy group, and in 3C 449 one finds $\beta \approx 30$ at the group radius, and $\beta \approx 400$ in the centre, showing that the magnetic field is likely to be dynamically insignificant in the central regions of groups and clusters.

Rotation measures have also been mapped to some extent outside of the central regions (Coma Cluster, A 119, A 514, A 400, A 2634, 3C 129). These data lead to estimates of the magnetic-field strength of $\sim$2 ... 8 μG, with ordered magnetic fields on scales of $\sim$5 ... 15 kpc. The variance σ_{RM} of the rotation measure as a function of distance from the cluster centre can be inferred from the distribution of the thermal gas density, assuming that the strength of the magnetic field obeys the same radial law. X-ray observations have revealed that the thermal gas density can be best described by

$$n_e(r) = n_0 \cdot \left(1 + \frac{r^2}{r_c^2}\right)^{-\frac{3\beta}{2}}, \tag{8.2}$$

where n_0 is the central density, r_c is the core radius, and β is a free parameter (not to be confused with the plasma-β!). Typical values are $r_c = 200$ kpc, $n_0 = 10^{-2}$ cm^{-3}, $\beta \approx \frac{2}{3}$. For a Gaussian distribution of rotation measures with $\langle RM \rangle = 0$, the variance is given by

$$\sigma_{RM}^2 = \langle RM^2 \rangle = (812)^2 \, L_c \int (n_e \, B_\parallel)^2 \, dr,$$

where L_c is the cell size, i.e. the size scale on which the magnetic field is tangled. Inserting $n_e(r)$ from (8.2), the following expression results:

$$\sigma_{RM}(r) = \frac{K \, B \, n_0 \, r_c^{\frac{1}{2}} \, L_c^{\frac{1}{2}}}{\left[1 + \left(\frac{r}{r_c}\right)^2\right]^{\frac{6\beta-1}{4}}} \cdot \sqrt{\frac{\Gamma(3\beta - 0.5)}{\Gamma(3\beta)}},$$

where $K = 624$ if the source lies completely beyond the cluster and $K = 441$ if the source is located half-way through the cluster.

8.4 Radio Relics

Relic sources are a class of diffuse radio sources located near the peripheries of galaxy clusters. What they have in common with radio haloes is the lack of any obvious radio galaxies that could be powering them. However, unlike haloes, they have an elongated or irregular shape, and they are strongly polarised. The elongation is tangential to the cluster circumference. One of the first such peripheral structures that was found is located at the south-western periphery of the Coma Cluster, and is called 1253+275. It marks the contact surface between the main cluster and a subcluster that is falling into the main one from the south-west (projected direction). The observed magnetic-field orientation is parallel to the (tangential) elongation of the diffuse source. Another prominent example is located in A 2256. This is the cluster that also possesses a prominent radio halo. Meanwhile, about 40 of clusters are known to possess radio relics.

One of the most spectacular structures of this kind has been discovered recently in the northern outskirts of the galaxy cluster CIZA J2242.8+5301 at a redshift of $z = 0.1921$. Figure 8.6 shows the pronounced gradient in spectral index across the 'sausage'-like structure (top panel) as well as the strong linear polarisation, with degrees of $p \approx 50 \ldots 60\,\%$. Its total length is 2 Mpc. A double radio relic in the merging galaxy cluster ZwCl 0008.8+5215 is shown in Fig. 8.7, which illustrates the opposite location of the relics, with the X-ray emission shown in colour, and the galaxy distribution as dashed contours. Note how both, the gas and the galaxy distribution have been stretched in the direction of the merging, with the relic structures marking the opposite locations of the shocks produced in this process. These are examples demonstrating diffusive shock acceleration also operates on

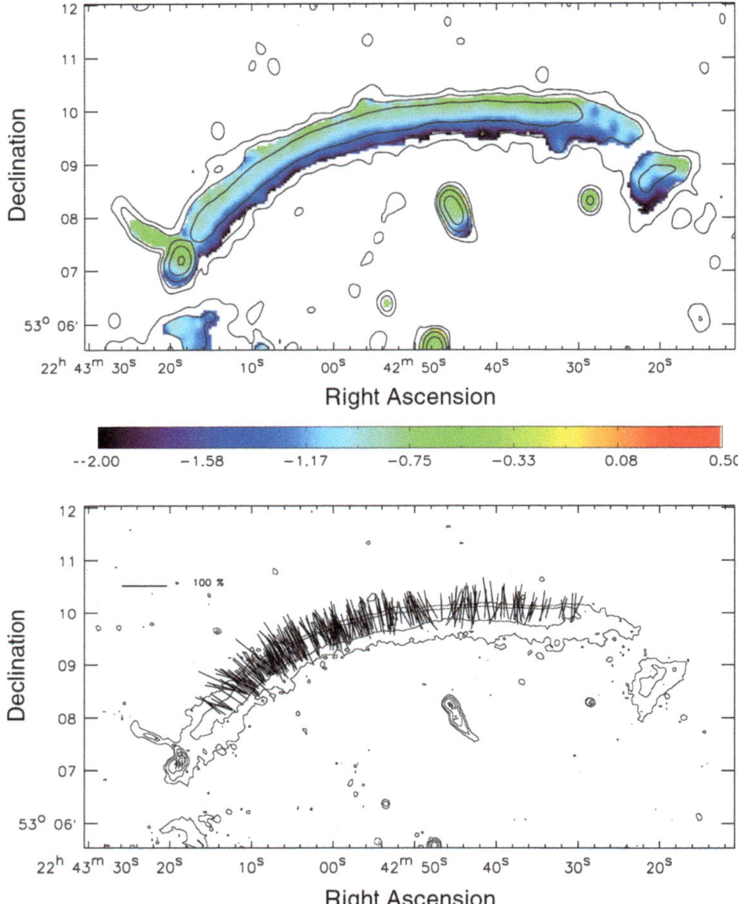

Fig. 8.6 Radio relic in the galaxy cluster CIZA J2242.8+5301 (van Weeren et al., 2010). The *upper image* shows the radio spectral index, superimposed onto the 20-cm map (contours). The *lower one* shows the electric-field vectors of the polarisation, with their length proportional the polarisation fraction

scales much larger than in supernova remnants and that shocks in galaxy clusters are capable of producing high-energy cosmic rays.

In summary, the properties of relics are:

- They have sizes of up to several Mpc.
- They are highly polarised, $p_{1.4\,\text{GHz}} \sim 10\dots50\,\%$.
- They possess steep radio spectra, $\alpha \geq 1.0$.
- Their locations are mostly peripheral.
- They are commonly found in merging clusters.

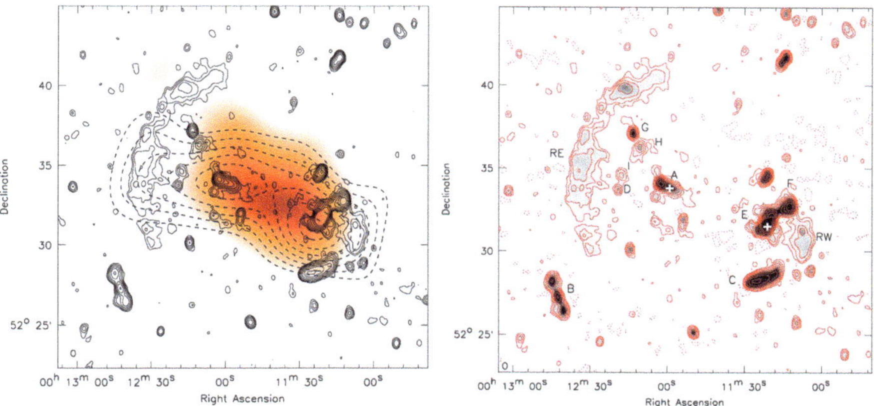

Fig. 8.7 Double relic in the galaxy cluster ZwCl 0008.8+5215 (van Weeren et al., 2011)

The most straight-forward explanation for this kind of diffuse synchrotron source in galaxy clusters is considered in the framework of the cluster-merging process in which large-scale shocks are produced:

(i) Fermi-I type, diffusive shock acceleration (DSA) of ICM electrons (the somewhat strange term 'radio gischt' has been coined for this phenomenon).
(ii) Adiabatic compression of a fossil radio plasma in cluster shocks. The plasma is the relic of mildly relativistic particles injected by AGN (or starburst galaxies) in the past (this phenomenon is referred to as 'phoenix').

The term 'radio gischt' stands for large sources. These cannot have been produced by compression, because the time scale for this compression would be too long to still see the particles radiating at GHz frequencies. The assumption here is that elongated the structures were equally extended in all directions prior to compression. The term 'phoenix' is used for smaller sources, for which the compression time scale is short enough to still see the relativistic particles radiating.

First analytical models of shock acceleration were able to reproduce the observed polarisation properties and spectral indices. In particular, the measured degree of linear polarisation is a good test, since the shock geometry predicts a dependence on the viewing angle of the relic structure. This relation has in fact been confirmed by observations. The predicted degree of linear polarisation as a function of viewing angle δ has been calculated by Enßlin et al. (1998) for weak fields (the magnetic pressure of the relic is small compared to the internal gas pressure) and strong fields (relic supported by magnetic pressure only):

$$\langle p(\delta) \rangle = \frac{g+1}{g+\frac{7}{3}} \cdot \frac{\sin^2 \delta}{F(R) - \sin^2 \delta} \,,$$

$$\nu_c = \frac{3}{4\pi} \cdot \frac{e \cdot B_\perp}{m_0 \, c} \cdot \gamma_{max}^2 \,,$$

where g is the power-law spectral index of the relativistic particles. The function $F(R)$ depends differently on the compression ratio R:

$$F(R) = \begin{cases} \frac{2}{15} \cdot \frac{13\,R-7}{R-1} & \text{strong magnetic fields} \\ \frac{2\,R^2}{R^2-1} & \text{weak magnetic fields} \end{cases} . \tag{8.3}$$

'Weak' and 'strong' here refer to the relative importance of the magnetic vs. the thermal pressure (plasma-β),

$$n\,k\,T \gtrless \frac{B^2}{8\,\pi} .$$

For the spectral models the examples shown in Figs. 8.6 and 8.7 represent templates of what may be nearly planar shocks seen edge-on. In this situation, the time elapsed since particles at a given location behind the shock were accelerated is $t = l/V_d$, where l is the distance from the shock front and V_d the shock downstream velocity. According to the DSA theory the injection spectral index α_{inj} is linked to the Mach number M of the shock by (e.g. Blandford and Eichler, 1987)

$$\alpha_{inj} = \frac{M^2 + 3}{2\,M^2 - 2} . \tag{8.4}$$

so the radio continuum spectra should exhibit a power-law of the form $I_\nu \propto \nu^{-\alpha_{inj}}$ directly behind the front of the shock. The synchrotron spectrum of the downstream regime can be calculated by making reasonable assumptions about the time evolution of the particle energy, as outlined in Sect. 7.4.3.

The observed radio continuum spectra appear to be power-law spectra from the lowest ($\sim$200 MHz) to the highest ($\sim$10 GHz) frequencies, with spectral indices from $\alpha \approx 0.8 \ldots 1.6$ and, perhaps surprisingly, with no indication for any spectral breaks. The question is whether one can find the break frequencies. If the observed spectra reflect the injection spectral indices, the straight power-laws render any such detection totally unlikely. If the break frequencies would be at low radio frequencies, there would not be 'much room' for the injection spectra, because the radio continuum of relics has been measured down to a few hundred MHz already. In addition, one should then see convex curvature in the spectra, which is not observed.

Such investigations now receive support by realistic model calculations. Three-dimensional MHD simulations of such passages of shock waves through a relic radio plasma have already been performed, with synthetic radio maps resulting from these (Enßlin and Brüggen, 2002; Skillman et al., 2013, see also Fig. 8.8). Any such MHD simulation reproducing the phenomenon of cluster radio haloes or relics is obviously a rather complex task and requires the implementation of MHD in large-scale cosmological simulations of hierarchial structure formation, while at the same time considering all physical processes that involve magnetic fields, particle acceleration, gas dynamics, and radiation. The observed radio emission is a product of the merger activity of galaxy clusters, and thus provides a powerful diagnostic

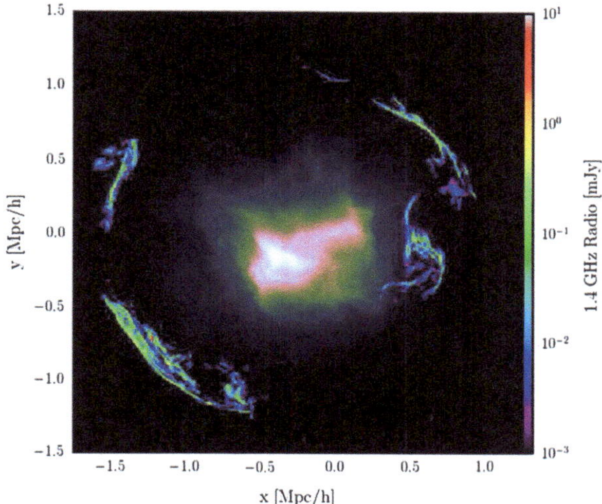

Fig. 8.8 Simulated radio and X-ray emission (Skillman et al., 2013)

tool to study galaxy-cluster merging from the local universe out to mild redshifts. The sensitivity of radio telescopes to these diffuse and partly faint structures is constantly improving with LOFAR, the VLA and the SKA.

8.5 Mini Haloes

The so-called mini-haloes in some cluster centres have a nature that is entirely different from the diffuse radio sources discussed in the preceding sections. As the name says, they are a lot smaller ($\leq 100\,$kpc) and are found around powerful radio galaxies residing at the centres of what are believed to be cooling-flow clusters. Cooling flows are inferred from the sharply peaked X-ray surface brightness distribution that is found in a large fraction of galaxy clusters. This peaked X-ray brightness is indicative of a cooling flow, because this reflects a steeply rising gas density towards the center of a cluster. The high central density in turn implies a short cooling time. The density of the piled-up gas has a high pressure

$$P = n\,k\,T$$

and thus strongly confines the radio plasma injected by a central AGN into its surroundings. A prototype of such a mini-halo is the extended radio emission around Per A (NGC 1275, 3C 84, see Fig. 8.10).

Other examples are Vir A (Fig. 8.9), PKS 0745-191, and possibly A 2390. Like other extended cluster sources, they also show steep radio spectra. The magnetic-field strength inferred from their (high) radio brightness is of order $B_t \approx 10\ldots20\,\mu$G. The radiative lifetimes of relativistic electrons in magnetic

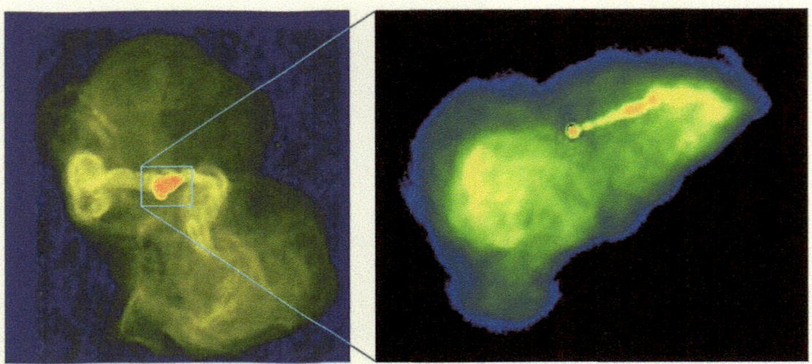

Fig. 8.9 The radio galaxy Vir A with its radio halo at $\lambda = 90\,\mathrm{cm}$ (*left*, Owen et al. 2000), and its inner lobes at $\lambda = 20\,\mathrm{cm}$ (Hines et al., 1989)

fields as strong as this are short: such particles would be observable for at most a few times 10^7 year – much too short to make it to the distances marked by the mini-haloes, away from the central radio galaxies. It has therefore been proposed that mini-haloes are powered by MHD turbulence in these cooling-flow centres. Gitti et al. (2004) have argued that the power necessary to accelerate the relic electron population is a mere $0.7\,\%$ of the maximum power that can be tapped from the cooling flow.

But, also here, there is an alternative explanation: the relativistic electrons could again be of secondary origin, being produced by hadronic collisions of the relativistic protons with the thermal ones.

8.6 Inverse-Compton Emission

A completely independent way of determining magnetic-field strengths would rest upon inverse-Compton X-ray emission, together with synchrotron radiation. Inverse-Compton emission is the relativistic extrapolation of the Sunyaev-Zel'dovich effect,[2] in which the ambient photon field is up-scattered into the hard X-ray regime by the same relativistic particles that produce the synchrotron radiation. This inverse-Compton process involves two Lorentz transformations (to and from the rest frame of the electron), plus Thompson scattering in the rest frame of the electrons, resulting in the emergent photon frequency

$$\nu_{IC} = \frac{4}{3} \cdot \gamma^2\, \nu_{CMB},$$

---

[2] The SZ effect implies scattering of CMB photons of the thermal free electrons of the hot plasma of galaxy clusters, leading to a decrement of the CMB brightness towards such clusters.

where ν_{CMB} is the frequency of the incident CMB photons, and γ the Lorentz factor of the relativistic electrons. Detecting the nonthermal hard X-ray and synchrotron radiation that is produced by the same population of relativistic electrons allows us to unambiguously estimate the volume-averaged strength of the intra-cluster magnetic-field. The exact derivations of the respective flux densities by Blumenthal and Gould (1970) yield the following expressions:

$$S_{syn}(\nu_r) = \frac{V'}{4\pi D_L^2} \cdot \frac{4\pi e^3}{(m_e c^2)^g} \cdot N_0 \, B^{\frac{g+1}{2}} \cdot \left(\frac{3e}{4z\pi m_e c} \right)^{\frac{g-1}{2}} \cdot A(g) \, \nu^{-\frac{g-1}{2}}$$

$$S_{IC}(\nu_x) = \frac{V'}{4\pi D_L^2} \cdot \frac{8\pi^2 a_0^2}{c^2} \cdot h^{-\frac{g+3}{2}} \, N_0 \, (m_e c^2)^{1-g} \cdot (k \, T_{CMB})^{\frac{g+5}{2}} \cdot F(g) \, \nu^{-\frac{g-1}{2}}.$$

Here:

$$
\begin{aligned}
a_0 &= \text{classical electron radius, } a_0 = e^2/m_e c^2 = 2.82 \cdot 10^{-13} \text{ cm} \\
h &= \text{Planck's constant, } h = 6.6262 \cdot 10^{-27} \text{ erg s} \\
g &= \text{power-law index of the relativistic electron spectrum} \\
V' &= \text{effective source volume, } V' = f \cdot V, \text{ where } f = \text{filling factor} \\
B &= \text{magnetic-field strength} \\
T_{CMB} &= \text{radiation temperature of CMB, } T = T_0 \cdot (1 + z) \\
N_0 &= \text{amplitude of the electron spectrum} \\
A(g), \, F(g) &= \text{tabulated functions given by Blumenthal and Gould (1970)} \\
D_L &= \text{luminosity distance of the cluster}
\end{aligned}
$$

$$B \propto (1+z) \left[\frac{S_{syn}(\nu_r)}{S_{IC}(\nu_x)} \right]^{\frac{2}{g+1}} \cdot \left(\frac{\nu_r}{\nu_x} \right)^{\frac{g-1}{g+1}}.$$

The CMB Planck function has its maximum at $\nu_{CMB} \approx 1.6 \cdot 10^{11}$ Hz, hence hard X-rays observed at 20 keV ($\nu_{IC} = 4.8 \cdot 10^{18}$ Hz) are produced by electrons with $\gamma = 5000$ (independent of redshift, since $\nu_{CMB} = \nu_{CMB,0} \cdot (1+z)$). The corresponding synchrotron emission of such relativistic electrons peaks at a (rest frame) frequency of

$$\nu_{syn} \approx 4 \cdot \left(\frac{B}{\mu G} \right) \cdot \gamma^2 \, \text{Hz} = 100 \cdot \left(\frac{B}{\mu G} \right) \, \text{MHz}.$$

The principal difficulty of measuring the inverse-Compton emission from galaxy clusters in the hard X-ray regime is the contamination or confusion by the thermal emission from the cluster atmosphere. One hence has to separate the thermal from the nonthermal (power-law) radiation through very sensitive spectroscopic X-ray measurements. These were rendered feasible with the Beppo/Sax and RXTE missions, allowing sensitive measurements well above 10 keV. This way, field strengths in the range $B \approx 0.2 \ldots 1 \, \mu G$ have been derived (Fusco-Femiano et al., 2004). These appear to be lower than those inferred from radio data alone. An interesting solution of this problem is to account for small pitch angles of the electrons, which would produce weaker synchrotron but still the same inverse-Compton emission!

8.7 Magnetisation of the IGM

The question of what injects particles and fields in galaxy clusters finally needs to be discussed. In view of the observed nonthermal phenomena, one might speculate whether or not galaxy clusters are filled with a (mildly) relativistic plasma throughout, which at least in view of the statistics of rotation measures (Fig. 8.4) seems to be the case. A primordial origin of magnetic fields, e.g. with the *Biermann battery* (Biermann 1950) is conceivable; however, the time scale to amplify them to the observed strengths is just too large, since there must be a mechanism that amplifies the magnetic fields from $\gtrsim 10^{-18}$ G to $\gtrsim 1\,\mu$G. It is therefore near at hand to consider injection of particles and fields by active galaxies. These could have been

- AGN
- Starburst (dwarf) galaxies

The injection of radio plasma into the ICM by AGN can still be seen, even in the local universe. Radio galaxies fill large volumes with the plasma, sometimes leaving behind pronounced cavities in the thermal plasma as evident in many X-ray observations. Prominent examples are Cyg A and Per A (Fig. 8.10, left). The central galaxies in the corresponding clusters have luminosities between $\sim 2\cdot 10^{38}$ and $7\cdot 10^{44}$ erg s^{-1}. The X-ray cavities have average radii of ~ 10 kpc and average projected distances of ~ 20 kpc from the central galaxy. The minimum energies associated with these activities are between $P\cdot V = 10^{55}$ erg (in galaxy groups) and $P\cdot V = 10^{60}$ erg (in rich clusters). The largest such structure was, however, found in the

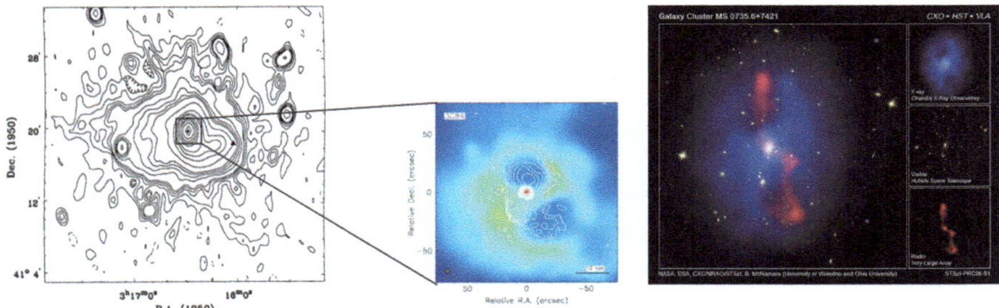

Fig. 8.10 Cavities produced by the radio plasma of AGN. *Left*: 327-MHz map of the mini-halo in the Perseus cluster produced by 3C 84, the blow-up showing the X-ray emission, with radio contours superimposed (From Gitti et al., 2004). *Right*: X-ray (*blue*) and radio (*red*) maps of MS 0735.6+7421 (McNamara and Nulsen, 2007)

$z = 0.22$ cluster MS 0735.6+7421 (Fig. 8.10, right). The cavities there measure $\sim$200 kpc in diameter, the inflation work is $P \cdot V = 10^{61}$ erg for each cavity. The enthalpy

$$H = U + PV,$$

where U is the internal energy, is much larger than that of other sources with prominent cavities:

$$H_{0735.6} \approx 10 \cdot H_{HydA},$$

$$\approx 15 \cdot H_{CygA},$$

$$\approx 250 \cdot H_{PerA},$$

$$\approx 10^4 \cdot H_{VirA}.$$

Another source of relativistic plasma can be seen in starburst galaxies, in particular low-mass ones. Kronberg et al. (1999) were the first to discuss possible seeding of magnetic fields (along with the ejection of relativistic particles) in the ICM by dwarf galaxies. This was treated more quantitatively, with predictions of magnetic-field strengths, by Bertone et al. (2006). Energetic winds from galaxies should have been able to provide seed fields with a strength of $B_t \approx 10^{-12} \ldots 10^{-8}$ G in the ICM, strong enough to be amplified to the observed values by the turbulent dynamo. Dwarf galaxies could have been responsible for (at least part of) the magnetisation of the ICM because of two advantages they bring:

1. They were/are numerous in the framework of the ΛCDM bottom-up scenario of structure formation.
2. Their shallow gravitational potentials imply low escape speeds,

$$v_{esc} = \sqrt{\frac{2\,G\,M}{R}}.$$

In judging their relative importance in magnetising the ICM one has to compare radio powers:

(i) Typical starburst dwarf galaxies produce $P_{1.4\,\mathrm{GHz}} \approx 10^{20.5}$ W Hz^{-1}.
(ii) Radio galaxies in the transition regime between FR I and FR II produce $P_{1.4\,\mathrm{GHz}} \approx 10^{24.7}$ W Hz^{-1},

i.e. $P_{FRI/II} \approx 15\,000 \cdot P_{SB}$. However, the bottom-up structure formation in a ΛCDM cosmology helps a lot here, since dwarf galaxies are much more numerous than massive spirals or ellipticals. This was all the more true in the early universe, which was a lot smaller then, helping to 'pollute' it with heavy elements, relativistic particles, and magnetic fields. Furthermore, the duty cycle of radio galaxies is probably low. By comparing the measured lengths of FR II radio galaxies with the projected length evolution of mock radio galaxies, Bird et al. (2008) estimate the lifetime of FR II radio galaxies in groups and clusters to be

$$\tau_{RG} \approx 1.5 \cdot 10^7 \text{ year}$$

and a duty cycle of

$$\tau_{DC} \approx 8 \cdot 10^8 \text{ year.}$$

The effective activity period can then be estimated via

$$\tau_{AP} = \frac{\tau_{univ}}{\tau_{RG} + \tau_{DC}} \cdot \tau_{RG},$$

where the ratio on the right-hand-side of this equation is the number of activity periods. This yields an effective activity period of

$$\tau_{AP} \approx 2 \cdot 10^8 \text{ year} \approx 0.018 \cdot t_{Hubble}.$$

As far as starburst dwarf galaxies are concerned there are local templates with current (e.g. NGC 4449), or recent (e.g. NGC 1569, see Sect. 5.3) starbursts. During the lifetime of the ejected relativistic particles these could propagate a considerable distance away from the galaxy, once they escaped from it. Outside the galaxy the energy losses are governed by inverse-Compton losses against the CMB photons. The existence of winds in low-mass galaxies is inferred from the observed kinematics of the gas (measured with slit spectroscopy), but can arguably be also inferred from measurements of the temperature of the hot (X-ray-emitting) gas. For instance, Martin (1998) found NGC 1569 to be a candidate galaxy in which the outflow speed exceeds the escape speed, and della Ceca et al. (1996) derived the temperature of its hot, X-ray-emitting gas to exceed the virial temperature for this galaxy.

Even though such particles age fairly rapidly so that they become invisible at GHz frequencies after $\sim 10^8$ year, the new low-frequency developments in radio astronomy provide ideal tools to trace them. With LOFAR working in two bands, the 'low band' (10–80 MHz) and the 'high band' (120–250 MHz), it will have the best low-frequency coverage to disclose the secrets of the nonthermal universe. It will thus be possible to trace relic haloes around all kinds of formerly active galaxies, also dwarf galaxies. Previously starbursting dwarf galaxies should be 'wrapped' in nonthermal radio haloes, detectable only at the lowest frequency.

If the relativistic electrons would propagate without 'intergalactic weather', i.e. just by diffusion, then they could maximally move with the Alfvén speed:

$$v_A = 2.2 \cdot \left(\frac{B}{\mu G} \right) \cdot \left(\frac{n_e}{\text{cm}^{-3}} \right)^{-\frac{1}{2}} \text{ km s}^{-1},$$

where n_e is thermal electron density. Taking $B = 1\,\mu G$ and $n_e = 0.001\,cm^{-3}$, the particles radiating at $120\,MHz$ could move out to $30\ldots40$ kpc within $500\,Myr$, or correspondingly further if caught at still lower frequencies. Hence, such haloes in which we should see starburst dwarf galaxies wrapped could have sizes of $60\ldots80$ kpc. LOFAR should be sensitive enough to detect such nonthermal haloes.

In case of 'stormy cluster weather', however, the situation will change dramatically. If located in a massive cluster undergoing merging, the ejected radio plasma would 'fly away' at high speeds, of order $1000\,km\,s^{-1}$. This is what one actually observes in the tails of the much more powerful WAT/NAT sources. Both, particles and fields would in this way spread out very quickly over the cluster volume, also taking into account the large number of such dwarf galaxies.

In any case, and doubtlessly, LOFAR will make a significant contribution to the scenarios discussed above. Even though we cannot trace the mechanisms back to the era of initial galaxy formation, we can test the hypotheses using observations in the local universe.

References

Bertone, S., Enßlin, Vogt, C., 2006, MNRAS **370**, 319
Biermann, L, 1950, Z. Naturforsch. **5a**, 65
Bird, J., Martini, P., Kaiser, C., 2008, Astroph. J. **676**, 147
Blandford, R., Eichler, D., 1987, Phys. Rep. **154**, 1
Blumenthal, G.R., Gould, R.J., 1970, Rev. Mod. Phys. **42**, 237
Bonafede, A., 2010, Ph.D. Thesis, Univ. Bologna
Clarke, T.E., Kronberg, P.P., Böhringer, H., 2001, Astroph. J. **547**, 111
Clarke, T.E., Enßlin, T.A., 2006, Astron. J. **131**, 2900
della Ceca, R., Griffiths, R.E., Heckman, T.M., MacKenty, J.W., 1996, Astroph. J. **469**, 662
Enßlin, T.A., Biermann, P.L., Klein, U., Kohle, S., 1998, Astron. Astroph. **332**, 395
Enßlin, T.A., Brüggen, M., 2002, MNRAS **331**, 1001
Feretti, L., Giovannini, G., Govoni, F., Murgia, M., 2012, Astron. Astroph. Rev. **20**, 54
Fusco-Femiano, R., Orlandini, M., Brunetti, G., et al., 2004, Astroph. J. **602**, 73
Garrington, S.T., Leahy, J.P., Conway, R.G., Laing, R.A., 1988, Nature **331**, 147
Gitti, M., Brunetti, G., Feretti, L., Setti, G., 2004, Astron. Astroph. **417**, 1
Hines, D.C. Eilek, J.A., Owen, F.N., 1989, Astroph. J. **347**, 713
Kronberg, P.P., Lesch, H., Hopp, U., 1999, Astroph. J. **511**, 56
Martin, C.L., 1998, Astroph. J. **506**, 222
McNamara, B.R., Nulsen, P.E.J., 2007, ARA&A **45**, 117
Owen, F.N., Eilek, J.A., Kassim, N.E., 2000, Astroph. J. **543**, 611
Skillman, S.W., Xu, H., Hallman, E.J., et al., 2013, Astroph. J. **765**, 215
Taylor, G.B., Perley, R.A., Inoue, M., et al., 1990, Astroph. J. **360**, 41
Taylor, G.B., Perley, R.A., 1993, Astroph. J. **416**, 554
van Weeren, R.J., Röttgering, H.J.A., Brüggen, M. Hoeft, M., 2010, Science **330**, 347
van Weeren, R.J., Hoeft, M., Röttgering, H.J.A., et al., 2011, Astron. Astroph. **528**, 38
van Weeren, R.J., Röttgering, H.J.A., Rafferty, D.A., et al., 2013, Astron. Astroph. **543**, 43
Willson, M.A.G., 1970, MNRAS **151**, 1

Chapter 9
Cosmological Magnetic Fields

Finally, a few words are mandatory about a cardinal question, the discussion of which can become very complex and is still a matter of much speculation. Whatever mechanism (in clusters, galaxies or on smaller scales) produces magnetic fields, there is the requirement of seed fields that have to be amplified. This holds especially for systems with large dynamical time scales (formation, rotation, turbulence). For instance, the sun has existed for some 10^{11} dynamical time scales, while the Milky Way is likely to have seen a mere 50 or so, which means that there were only 50 e-folding times to build up the magnetic field. Even worse: strong magnetic fields are already present at large redshifts. Cosmological magnetic fields have been discussed to some extent in the excellent review by Widrow (2002). One mechanism that has been briefly addressed in the previous chapter is the Biermann battery Biermann (1950).

Explaining the Biermann battery effect starts out from the difference between the mobility of electrons and ions in a plasma, which we have to account for via the two momentum equations (see Widrow, 2002),

$$\frac{d\vec{V}_e}{dt} = -\frac{\vec{\nabla}p_e}{\rho_e} - \frac{e}{m_e}\left(\vec{E} + \frac{\vec{V}_e \times \vec{B}}{c}\right) - \vec{\nabla}\Phi + \frac{\vec{K}_{ep}}{m_e}, \tag{9.1}$$

$$\frac{d\vec{V}_p}{dt} = -\frac{\vec{\nabla}p_p}{\rho_p} - \frac{e}{m_p}\left(\vec{E} + \frac{\vec{V}_p \times \vec{B}}{c}\right) - \vec{\nabla}\Phi + \frac{\vec{K}_{ep}}{m_p}. \tag{9.2}$$

Here, p_i is the partial pressure, m_i the mass, $\rho_i = m_i n_i$ the mass density, n_i the number density, and $\vec{V}_i$ the velocity of the electrons $i = e$ and protons $i = p$. $\vec{K}_{ep} = m_e(\vec{V}_p - \vec{V}_e)/\tau$ describes the momentum transfer from protons to electrons, where the characteristic time scale of electron-proton collisions is $\tau = m_e\sigma/n_e e^2$ for a conductivity σ. If local charge neutrality holds, then with a current $\vec{J} = e(n_p\vec{V}_p - n_e\vec{V}_e)$ the momentum transfer becomes $\vec{K}_{ep} \approx e\vec{J}/\sigma$. Subtracting Eqs. (9.1) and (9.2), one obtains the generalised Ohm's law,

© Springer International Publishing Switzerland 2015
U. Klein, A. Fletcher, *Galactic and Intergalactic Magnetic Fields*,
Springer Praxis Books, DOI 10.1007/978-3-319-08942-3_9

$$\frac{m_e}{e^2} \frac{\partial}{\partial t} \left(\frac{\vec{J}}{n_e} \right) = \frac{m_e}{e} \frac{d(\vec{V}_p - \vec{V}_e)}{dt} \tag{9.3}$$

$$= \frac{\vec{\nabla} p_e}{e\, n_e} + \vec{E} + \frac{\vec{J} \times \vec{B}}{c\, n_e} + \frac{\vec{V}_p \times \vec{B}}{c} - \frac{\vec{J}}{\sigma} . \tag{9.4}$$

The term describing the gravitational force has disappeared, since the curl of $\vec{\nabla}\Phi$ vanishes. Terms of order m_e/m_p have been neglected, and it is assumed that the partial pressures of the electrons and protons are identical. The $\vec{J} \times \vec{B}$ term can be ignored for the presentation consideration, as we are only interested in the creation of weak seed fields. For a fluid with velocity $\vec{V}_p$,

$$\vec{J} = \sigma \left(\vec{E} + \frac{\vec{V}_p \times \vec{B}}{c} \right)$$

is exactly Ohm's law so that Eq. (9.4) now reads

$$\frac{m_e}{e} \frac{\partial}{\partial t} \left(\frac{\vec{J}}{n_e} \right) = \frac{\vec{\nabla} p}{n_e} + e\, \vec{E} . \tag{9.5}$$

The curl of this is

$$\frac{m_e}{e} \vec{\nabla} \times \frac{\partial}{\partial t} \left(\frac{\vec{J}}{n_e} \right) = - \frac{\vec{\nabla} n_e \times \vec{\nabla} p}{n_e^2} + e\, \vec{\nabla} \times \vec{E} , \tag{9.6}$$

which, together with Maxwell's equations, yields

$$\frac{\partial \vec{B}}{\partial t} - \vec{\nabla} \times (\vec{V} \times \vec{B}) = \frac{m_e\, c}{e} \cdot \frac{\vec{\nabla} p_e \times \vec{\nabla} \rho_e}{\rho_e^2} . \tag{9.7}$$

In a real fluid, the pressure and density gradients are not necessarily parallel to each other (there may be curved shocks and photoheating) so that the term on the right-hand side of Eq. (9.7) can be non-zero! This extra term is due to gas-dynamical effects and thus produces a seed magnetic field once there is vorticity in the fluid.

Various authors have thought about magnetic-field creation in exotic, ultra-dense stages of the Big Bang, by considering what might have happened during phase transitions:

- Electro-weak era: 10^{-12} s.
- Quark-hadron era: 10^{-5} s; the event horizon then encompassed $10^6\, M_\odot$ of baryons, hence the resulting field on galactic scales would be only 10^{-30} G.
- GUT era: $10^{-36} \ldots 10^{-34}$ s, 10^{15} GeV; the horizon then contained a mere 10^4 baryons; even with a high local energy density, there was no chance to get to 10^{-20} G on galactic scales.

Other scenarios deal with proto-galactic batteries, such as the Compton drag: since the cross section for Inverse-Compton scattering is much larger for electrons than for protons, the former lose energy against the photon background, this changing their bulk motion with respect to the latter. This produces a current, which in turn produces a magnetic field of 10^{-21} G at $z = 5$.

Another ingenious idea (Harrison, 1970) rests upon the speculation whether there has been vorticity in the primordial fluctuations. If so – and nobody really knows – there would have been an interesting mechanism leading to a proton-electron separation. The main criticism of this hypothesis is the lack of vorticity, which can not be created by gravity alone. What is required is gas dynamics, which creates turbulence and tidal forces, which in turn can give rise to curl. Now let us consider Harrison's mechanism.

Consider a spherical region of radius r in the pre-recombination era, which consisted largely of a soup of photons, protons, and electrons (plus dark matter, neutrinos,...). The electrons were tightly coupled to the photons by Thomson scattering, while the coupling to the ions was weaker. The sphere has photons with uniform density ρ_γ and matter with uniform density ρ_m. As the eddy expands, we have

$$\rho_m \, r^3 = const$$

and

$$\rho_\gamma \, r^4 = const.$$

If ω_γ and ω_m denote the angular velocities of the photon and matter spheres, respectively, then the angular momenta

$$L_\gamma = \rho_\gamma \, \omega_\gamma \, r^5$$

and

$$L_m = \rho_m \, \omega_m \, r^5,$$

are separately conserved. Hence we find

$$\omega_\gamma \propto r^{-1} \tag{9.8}$$

and

$$\omega_m \propto r^{-2}, \tag{9.9}$$

because the eddy expands with the cosmic scale factor. We thus see that radiation spins down more slowly than matter. Because of the coupling of the electrons and photons by Thomson scattering, the electron soup will rotate faster than the ion soup. The difference in angular velocities leads to a magnetic field

$$\vec{B} = -2 \cdot \frac{m_H\, c}{e} \cdot \vec{\omega} = -2.1 \cdot 10^{-4} \cdot \vec{\omega}\ \text{G}.$$

Another mechanism that may induce cosmological magnetic fields by counter-streaming of electron-ion plasmas is the so-called Weibel instability (Schlickeiser, 2005). Such streaming motions can be induced in the course of structure formation in the early universe. This process requires large Mach numbers ($M > 43$) of the fully ionised baryonic flows. There may be other scenarios, but the intrigued reader is referred here to specific literature (see the comprehensive review by Widrow 2002).

References

Biermann, L, 1950, Z. Naturforsch. **5a**, 65
Harrison, E.R., 1970, MNRAS **147**, 279
Widrow, L.M., 2002, Rev. Mod. Ph. **74**, 775
Schlickeiser, R., 2005, Plasm. Phys. Contr. Fusion **47**, A205

Index

© Springer International Publishing Switzerland 2015
U. Klein, A. Fletcher, *Galactic and Intergalactic Magnetic Fields*,
Springer Praxis Books, DOI 10.1007/978-3-319-08942-3